Impact of process conditions in open and closed reactor systems on the properties of thermally modified wood

Impact of process conditions in open and closed reactor systems on the properties of thermally modified wood

Dissertation

in partial fulfillment of the requirements for the doctoral degree (Dr. rer. nat.)
of the Faculty of Forest Sciences and Forest Ecology
Georg-August University Göttingen

within the PhD program "Wood Biology and Wood Technology" of the Georg-August University School of Science (GAUSS)

Submitted by

Michael Altgen

Born in Mechernich, Germany

Göttingen, 2016

Bibliografische Information der Deutschen Nationalbibliothek

Die Deutsche Nationalbibliothek verzeichnet diese Publikation in der Deutschen Nationalbibliografie; detaillierte bibliografische Daten sind im Internet über http://dnb.d-nb.de abrufbar.

1. Aufl. - Göttingen: Cuvillier, 2016
   Zugl.: Göttingen, Univ., Diss., 2016

   ISBN 978-3-7369-9421-8
   eISBN 978-3-7369-8421-9

Members of the examination board

Referee: Prof. Dr. Holger Militz, Wood Biology and Wood Products, Burckhardt Institute, Faculty of Forest Sciences and Forest Ecology, Georg-August University, Göttingen, Germany

Co-referee: Prof. Dr. Carsten Mai, Wood Biology and Wood Products, Burckhardt Institute, Faculty of Forest Sciences and Forest Ecology, Georg-August University, Göttingen, Germany

2nd Co-referee: Prof. Dr. Andreas Krause, Institute of Mechanical Wood Technology, Department of Wood Sciences, University of Hamburg, Germany

Further members of the examination board

Prof. Dr. Ursula Kües, Molecular Wood Biotechnology and Technical Mycology, Büsgen Institute, Faculty of Forest Sciences and Forest Ecology, Georg-August University, Göttingen, Germany

Prof. Dr. Stefan Schütz, Forest Zoology and Forest Conservation, Büsgen Institute, Faculty of Forest Sciences and Forest Ecology, Georg-August University, Göttingen, Germany

Jr-Prof. Dr. Kai Zhang, Wood Technology and Wood Chemistry, Burckhardt Institute, Faculty of Forest Sciences and Forest Ecology, Georg-August University, Göttingen, Germany

Date of oral examination: 18th November 2016

Acknowledgement

First and foremost, I thank Prof. Dr. Holger Militz, my supervisior, for giving me the opportunity to work on my dissertation at his department. I greatly appreciate his support, encouragement, patience and expertice during the entire study.

This work could not have been completed without my friends and colleagues at the department Wood Biology and Wood Products as well as all the students who helped me in planning, preparing and carrying out my experiments. I especially owe my thanks to the members of the "Buena Vista Paper Club", André Klüppel, Bennedikt Hünnekens, Bernd Lütkemeier, Daniela Altgen and Kim Krause for giving me constructive feedback on my manuscripts and for joining in on entertaining lunch-breaks.

Cordial thanks go to my co-authors Prof. Dr. Stergios Adamopoulos, Dr. habil Tamás Hofmann and Dr. Wim Willems. I am very grateful to Wim for the many inspriring discussions we had in the past years and for his high commitment to get the reactor system running, which became an integral part of my research work.

I would also like to thank Jukka Ala-Viikari and Timo Tetri from the International ThermoWood Asscocation, as well as Janne Pynnonen and Duncan Mayes from Stora Enso Wood Products for their financial support and fruitful discussions throughout my studies.

This thesis is dedicated to my wife Daniela and my parents for their everlasting encouragement and support, for uplifting me when things went wrong and sharing my happiness when things worked out.

Abstract

This thesis investigates different process conditions of thermal wood modification technologies in open and closed reactor systems and their influence on the resulting wood properties.

The first part of this thesis studies thermal modification processes in closed reactor systems. Thermal modification of wood in closed reactor systems allows for the control of the water vapor pressure, which has a profound impact on the thermal degradation and the change in wood properties. During the analysis, the accumulation of thermal degradation products in wood treated in high-pressure environments had to be considered as they resulted in an additional weight and a cell wall bulking effect. Wood degradation, quantified by the wood mass loss, increased with the maximum water vapor pressure irrespective of the temperature (150-180 °C) applied. As potential reasons, the improved heat transfer in a compressed gas atmosphere, reduced evaporative cooling, the accumulation of organic acids and the presence of water in the wood during the process were discussed. Furthermore, it was suggested that the wood moisture content affects chemical reactions in the wood polymeric network during the process, with preferential hydrolytic cleavage of bonds in wood treated in wet state, and enhanced bond formation via esterification, condensation and cross-linking reactions in wood treated in dry state. For the reduction in hygroscopicity of wood by thermal modification, three mechanisms were identified: (1) The reduction in the number of sorption sites for water by the preferential degradation of hemicelluloses. (2) The increase in the cell wall matrix stiffness by ultra-structural realignments of amorphous cell wall matrix polymers caused by drying and softening of wood at elevated temperatures. However, this effect was shown to be reversible by conditioning at high humidity or water-soaking. (3) The irreversible increase in the cell wall matrix stiffness by the modification and the formation of covalent bonds within the lignin carbohydrate complex during thermal modification processes of wood in dry-state. Some of these mechanisms were also shown to influence other wood properties, i.e. inelastic toughness (2) and maximum swelling (3). It was further demonstrated that the removal of hemicelluloses, without simultaneous modifications of the lignin carbohydrate complex does not result in a decrease, but in an increase in maximum swelling. The water vapor pressure applied during the process influences all three mechanisms and is thus considered as an additional control parameter to further optimize thermal modification processes.

The second part of the thesis investigates the surface performance of thermally modified Norways spruce and Scots pine wood from open reactor systems with respect to the impact of abiotic factors involved in weathering, as well as to the coatability with commercial coating systems. It was demonstrated that thermally modified wood is strongly susceptible to surface cracking during repeated capillary wetting and re-drying. This was explained by the presence of micro-cracks in thermally modified wood that act as starting points for more severe cracks and a reduced distribution of stresses generated by wetting and re-drying as a consequence of hemicelluloses removal and decreased cell wall matrix polymer mobility.

Wood defects caused by thermal modification were also related to an excessive penetration of a solvent-borne oil into thermally modified pine and spruce, since defects create more flow pathways. When exposed to UV irradiation and water-spray, thermal modification of wood did not prevent the photodegradation of the lignin. However, the rate at which micro-tensile strength loss occurred by photodegradation was lower for thermally modified micro-veneers, which was assigned to the pre-degradation of amorphous carbohydrates during the modification process. It was further demonstrated that despite a strong decrease in the wettability of thermally modified wood by water, a sufficient adhesion of water-borne coating systems can be achieved. However, the adhesion strength measured depended strongly on the mechanical interaction of the specific substrate/coating system. A brittle coating on thermally modified wood as a brittle substrate resulted in a very low adherence. Moreover, the presence of remaining native extractives and the accumulation of degradation products were considered as a risk for causing discolorations of coatings or for interfering with the curing and hardening reactions of coatings.

Zusammenfassung

Die vorliegende Doktorarbeit untersucht verschiedene Prozessbedingungen während der thermischen Holzmodifizierung in offenen und geschlossenen Reaktorsystemen und deren Einfluss auf die resultierenden Holzeigenschaften.

Der erste Teil dieser Doktorarbeit befasste sich mit der thermischen Holzmodifizierung in geschlossenen Reaktorsystemen unter Anwendung erhöhten Dampfdrucks. Während der Analyse mussten die im Holz verbleibenden Abbauprodukte berücksichtigt werden, da diese durch ihr zusätzliches Gewicht und die Blockierung von Zellwandporen Einfluss auf die Ergebnisse nahmen. Die Degradation des Holzes, quantifiziert durch den Masseverlust, stieg unabhängig von der angewendeten Temperatur (150-180 °C) mit dem maximalen Dampfdruck an. Als mögliche Ursachen wurden die verbesserte Wärmeübertragung in komprimierter Gasatmosphäre, die reduzierte Verdunstungskühlung, die Akkumulierung von organischen Säuren, sowie die erhöhte Holzfeuchte während des Prozesses diskutiert. Darüber hinaus deuteten die Ergebnisse darauf hin, dass die thermische Modifizierung feuchten Holzes zu vermehrter hydrolytischer Spaltung von Bindungen führt, während die Modifizierung trockenen Holzes möglicherweise die Bildung neuer Bindungen im Holz über Veresterung und Kondensationsreaktionen begünstigt. Für die Verringerung der Hygroskopizität des Holzes durch thermische Modifizierung wurden drei Mechanismen identifiziert: (1) Die Reduzierung der Anzahl an Sorptionsstellen für Wasser durch den bevorzugten Abbau von Hemicellulosen. (2) Der Anstieg der Zellwandmatrixsteifigkeit durch strukturelle Umorientierung amorpher Zellwandpolymere infolge der Trocknung und Erweichung von Holz bei erhöhten Temperaturen. Dieser Effekt war allerdings durch Konditionierung des modifizierten Holzes bei hohen Luftfeuchten oder durch Wasserlagerung rückgängig zu machen. (3) Der irreversible Anstieg der Zellwandmatrixsteifigkeit durch die Bildung zusätzlicher kovalenter Bindungen im Lignin-Kohlenhydrat-Komplex während der thermischen Modifizierung trockenen Holzes. Es zeigte sich, dass einige dieser Mechanismen auch andere Holzeigenschaften, wie den nicht-elastischen Anteil der Zähigkeit (2) und die maximale Quellung (3), beeinflussen. Der maximale Dampfdruck beeinflusst alle drei genannten Mechanismen und wird daher als zusätzlicher Kontrollparameter für die weitere Optimierung thermischer Modifizierungsprozesse angesehen.

Der zweite Teil dieser Doktorarbeit untersuchte das Oberflächenverhalten von thermisch modifiziertem Kiefern- und Fichtenholz aus offenen Reaktorsystemen hinsichtlich des Einflusses witterungsbedingter, abiotischer Faktoren, sowie der Verwendbarkeit kommerzieller Lacksysteme. Thermisch modifiziertes Holz zeigte sich während wiederholter Auffeuchtung und Abtrocknung sehr anfällig gegenüber der Bildung von Oberflächenrissen. Als Ursachen wurden das Vorhandensein von mikroskopischen Defekten in thermisch modifiziertem Holz als Startpunkt für makroskopische Risse, sowie die verminderte Spannungsübertragung im Holz aufgrund des Abbaus der Hemicellulosen und der Verminderung der Polymermobilität innerhalb der Zellwandmatrix diskutiert.

Mikroskopische Defekte in thermisch modifiziertem Holz als zusätzliche Eindringwege wurden außerdem mit einer erhöhten Eindringung eines lösemittelbasierten Öls in Verbindung gebracht. Während der Exposition des Holzes mit UV-Einstrahlung und Sprühwasser wurde kein Effekt der thermischen Modifizierung auf den Ligninabbau festgestellt. Zugversuche an Mikrofurnieren zeigten jedoch, dass der durch UV-Abbau bedingte Festigkeitsverlust für thermisch modifiziertes Holz weniger stark ausgeprägt war. Dies wurde durch die bereits im Modifizierungsprozess erfolgte Degradation der Hemicellulosen erklärt. Desweiteren wurde gezeigt, dass eine ausreichende Haftung von wasserbasierten Lacksystemen auf thermisch modifiziertem Holz trotz einer verschlechterten Benetzbarkeit mit Wasser möglich ist. Allerdings war die Haftung abhängig vom verwendeten Lacksystem. Die Verwendung eines spröden Lacksystems auf thermisch modifiziertem Holz als sprödem Substrat resultierte in einem deutlichen Rückgang der Haftung. Zudem wurden die nach dem Prozess im Holz verbleibenden nativen Extraktstoffe und Abbauprodukte, wie organische Säuren, als Risiko für die Verfärbung von Lacksystemen und für die Beeiflussung der Aushärtungsreaktionen von Lacksystemen angesehen.

Table of contents

Chapter 1   General introduction

1.1   Wood as a raw material

Wood serves as the main construction material for trees and allows for the capillary water transport between roots and leafs. It is formed by the vascular cambium and has a cellular structure with an interconnected network of macroscopic intracellular (lumen) and intercellular spaces (Fromm 2013, pp. 2). The cell walls are constructed as a complex biopolymer composite that form a material with high mechanical strength at low weight. Due to the unique properties of wood, it is regarded as one of the most important renewable resources for meeting the growing demand for bioenergy, construction materials and pulp for paper production. Within the UNECE region the consumption of roundwood has increased to 1.26 billion cubic meters in 2014 (UNECE/FAO 2015). The efforts for utilizing renewable raw materials that can be derived from sustainable sources and that can act as carbon sinks are likely to increase the demand for wood products (FAO 2012). In order to prevent shortage of supply and increasing raw material prices, an efficient material utilization is required. Besides recycling and material savings, the extension of the service of wood products is a key aspect.

1.1.1   Wood composition and ultra-structure

On an ultra-structural level, wood can be considered as a biopolymer composite built up of an interconnected network of cellulose, hemicelluloses and lignin with small amounts of extractives and inorganics. Based on its dry weight, wood is mainly composed of polysaccharides (65-75 %) and lignin (18-35%). The polysaccharides consist mainly of cellulose (40-45%) and hemicelluloses (15-25%), as well as of small proportions of starch and pectin (Rowell et al. 2005). The chemical composition differs between the wood species. In general, softwood species contain larger amounts of lignin, but lower amounts of hemicelluloses than hardwood species. Furthermore, the chemical compositions of lignin and hemicelluloses differ in softwoods and hardwoods, while cellulose is a uniform component in all wood species (Fengel and Wegener 1984 pp. 26). The three main polymers of the wood cell wall are briefly described below.

Cellulose is a linear high-molecular weight polymer consisting exclusively of D-glucopyranose units that are linked together by ß-(1→4)-glucosidic bonds. The repeating unit of the cellulose chain is a cellobiose unit, in which one of the two glucose units is turned around its C1-C4 axis of the pyranose ring. The cellulose chain has a reducing (C1-OH) and a non-reducing end (C4-OH) (Fengel and Wegener 1984 pp. 67). The size of the chain is usually specified as degree of polymerization (DP), which is the molecular weight of the cellulose chain related to the molecular weight of one glucose unit. For wood cellulose, an average DP of at least 9,000-10,000 is expected. Within the wood cell wall, cellulose is in a semi-crystalline state. Cellulose molecules have a tendency to form intra- and intermolecular hydrogen bonds and aggregate to microfibrils. These microfibrils contain highly ordered crystalline regions with a high packing density, as well as less ordered non-crystalline regions with a lower packing density. Up to 65 % of the wood cellulose exists as crystalline cellulose (Rowell et al. 2005).

Hemicelluloses are a composition of various sugar units, which may be pentose or hexose sugars. Compared to cellulose, hemicelluloses have much shorter, branched molecular chains and contain acetyl- and methyl-substituted groups (Fengel and Wegener 1984 pp. 106; Rowell et al. 2005). The most common sugar units in hemicelluloses are D-glucopyranose, D-galactopyranose, L-arabinofuranose, D-mannopyranose, D-glucopyranosyluronic acid and D-galactopyranosyuronic acid (Rowell et al. 2005). Softwood species usually have a high proportion of mannose units and more galactose units compared to hardwood species, which have a high proportion of xylose units and contain more acetyl groups (Fengel and Wegener 1984, p. 108).

Lignin is an amorphous, mainly aromatic polymer that consists of phenyl-propane units. These units are linked together to a three-dimensional polymer by C-O-C and C-C linkages. The building units and primary precursors of lignin are p-coumaryl alcohol, coniferyl alcohol and sinapyl alcohol. They have several reactive sides and their formation to a lignin macromolecule occurs through a random coupling to a non-linear polymer, rather than by a regular mechanism that is genetically prescribed (Fengel and Wegener 1984, pp. 132). Guaiacyl lignin, a polymerization product of guaiacyl alcohol, dominates the softwood lignins, while hardwood lignins are mainly syringyl-guaiacyl lignin, which is a copolymer of coniferyl and sinapyl alcohol (Rowell et al. 2005).

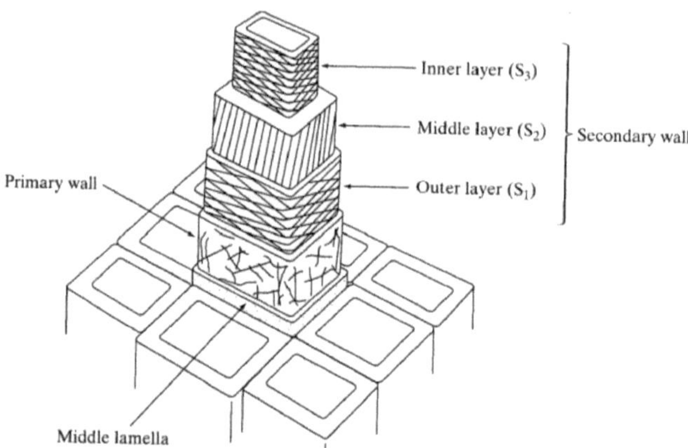

Figure 1: Cell wall structure of wood (derived from Dinwoodie 1989).

The three main polymers are not distributed evenly over the different cell wall layers of wood (**Figure 1**). The middle lamella and primary wall are mainly built up by lignin and contain much lower amounts of hemicelluloses and cellulose. However, the lignin content of the cell wall decreases towards the lumen, from the first layer (S$_1$ layer) to the second (S$_2$

layer) and third layer (S_3 layer) of the secondary wall. However, the bulk of all three main cell wall polymers is found in the S_2 layer, since this is the by far thickest layer of the cell wall (Fengel and Wegener 1984 pp. 227; Rowell et al. 2005). It makes up for approximately 80 % of the cell wall by weight and thus has a dominating influence on the wood properties, such as mechanical performance or swelling (Skaar 1988; Salmén and Burgert 2009). More than 50 % of the S_2 layer consists of cellulose (Fengel and Wegener 1984 p. 228; Rowell et al. 2005).

The arrangement of the polymers within the secondary cell wall shows some degree of order (ultra-structure), rather than being a random mixture. The cellulose microfibrils are imbedded in a matrix of amorphous hemicelluloses and lignin. The microfibrils have different inclinations with respect to the cell axis (microfibril angle, MFA) in the different layers. While the MFA is very high in the S_1 and S_3 layers, it is low in the S_2 layer and the microfibrils are almost oriented longitudinally (Rowell et al. 2005). The cellulose microfibrils are approximately 3-4 nm in diameter (Larsson et al. 1997; Wickholm et al. 1998). In the secondary cell wall, they are arranged in an undulating aggregate structure with aggregates reaching up to 30 nm in diameter. The spacing between the aggregates is lens shaped and contains amorphous cell wall polymers, hemicelluloses and lignin (Boyd 1982; Salmén and Burgert 2009; Salmén 2015). Hemicelluloses show a much lower degree of orientation than cellulose (Salmén et al. 2012). Hemicelluloses do not bond to the cellulose fibril surfaces covalently, but mainly by hydrogen bonds, which might be further supported by physical entanglement within the non-crystalline regions (Whistler and Chen 1991; Salmén and Olsson 1998; Salmén and Burgert 2009). In contrast, lignin is not bound to cellulose directly, but there are covalent bonds, i.e. ester, ether and glycosic bonds, between lignin and hemicelluloses that form lignin carbohydrate complexes that are difficult to separate (Whistler and Chen 1991; Koshijima and Watanabe 2003; Du et al. 2013). Hemicelluloses thus play a major role in maintaining the cell wall assembly (Salmén and Burgert 2009).

1.1.2 Natural properties of wood

The properties of wood are a result of a broad number of factors. On a macroscopic and microscopic level, factors such as the grain angle, the fiber length, the distribution of early- and latewood and various growth characteristics influence the mechanical properties of wood strongly (Winandy and Rowell 2005). There are also many factors that explain the differences in the properties of the various wood species. Nevertheless, the most fundamental aspects of wood properties can be explained by the general chemical composition and ultra-structure of the wood cell wall, mostly irrespective of the individual wood species.

Mechanical properties

Clear wood (free of knots, defects or irregular growth characteristics) has a high specific strength when compared to other building materials (Askeland et al. 2010, pp. 702), which can be linked to its unique structure and composition. When an external load is applied, the cell wall constituents contribute in different degrees to the strength of wood. The cellulose microfibrils function as tensile reinforcement of the cell wall. The tensile strength of cellulose is strongly dependent on its DP, with a decrease in DP leading to a reduction in strength (Ifju 1964). They exhibit an extremely high modulus of elasticity (MOE) under tension and thus contribute greatly to the stiffness of wood (Eichhorn and Young 2001). In compression, however, cellulose microfibrils buckle easily in contrast to their high strength and stiffness in tension (Gindl et al. 2004b). This low resistance against compressive loads of cellulose microfibrils is compensated by the rigid lignin matrix in which the microfibrils are imbedded. Lignin also limits the access of water to the cell wall and thereby contributes to the ability of wood to retain its strength and stiffness when exposed to moisture (Lagergren et al. 1957; Klüppel and Mai 2012). Since there are no direct bonds between cellulose and lignin, the hemicelluloses act as coupling agent between the cellulose microfibrils and the lignin matrix, thereby enabling the transfer of stresses between the individual cell wall polymers (Sweet and Winandy 1999; Winandy and Lebow 2001; Winandy and Rowell 2005).

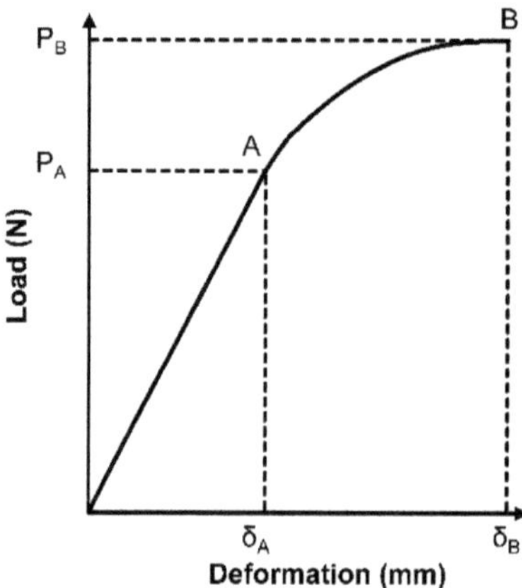

Figure 2: Schematic load-deformation curve showing the proportional limit (A) with its corresponding deformation (P_A) and load (δ_A) as well as the ultimate strength (B) with its corresponding load (P_B) and deformation (δ_B).

When load and deformation are plotted against each other, the result for wood materials is a load-deformation curve as shown in **Figure 2**. The test methods that can be used differ considerably in the way the external load is applied, leading to different types of stress within the specimen. The three primary types of stress are tensile, compressive and shear stress. When not considering creep and focusing on the immediate deformation, the behavior of wood under external load is characterized by a linear, elastic behavior below the proportional limit, and a non-linear, inelastic behavior above the proportional limit (see **Figure 2**). Below the proportional limit, all deformations are recoverable. The elastic deformation that occurs is caused by the breakage of hydrogen bonds between and within individual polymer chains, sliding of polymers by one another and subsequently the reformation of hydrogen bonds. Additionally, covalent bonds are distorting within the ring structures of polysaccharide chains. When exceeding the proportional limit, the stresses can no longer be distributed in a linear elastic manner and become great enough to cause covalent bond scission, which results in a permanent deformation. This non-linearity is associated with stress-induced plastic flow, during which the material is undergoing a rearrangement of its internal molecular and microscopic structure. In approaching the ultimate strength (see **Figure 2**), the elastic behavior becomes less notable, while failure and disorientation of microfibrils, the separation of the various cell wall layers, and the distortion of wood cells in relation to one another and finally failure of the wood cells by either scission of the cell, or cell-to-cell withdrawal occur (Winandy and Rowell 2005).

Moisture sorption and swelling of wood

Wood responds to variations in temperature and relative humidity with changes in moisture content. When temperature and relative humidity remain constant, wood will eventually reach a stable equilibrium moisture content. Wood reaches the fiber saturation point (FSP) at equilibrium conditions close to 100% relative humidity (RH), which corresponds to about 40% moisture content (Hoffmeyer et al. 2011). Above FSP, free water can be found in the cell lumen, whereas only bound water within the cell wall is present during water vapor sorption below the FSP (Zelinka et al. 2012; Engelund et al. 2013). Hydroxyl (OH) groups of the wood polymers are the main sorption sites for bound water. Water-accessible OH groups are mainly found in the hemicelluloses of the wood cell wall. In contrast, lignin has generally less OH groups compared to cell wall polysaccharides and cellulosic OH groups are only water-accessible on the surface of the microfibrils (Runkel 1954; Runkel and Lüthgens 1956).

When plotting the EMC against the relative humidity at a constant temperature (<75 °C), the result is a sorption isotherm with a hysteresis between adsorption (uptake) and desorption (release of water molecules) (Seborg and Stamm 1931; Stamm and Woodruff 1941), as shown in **Figure 3**. Typically, sorption isotherms of wood have a sigmoidal shape with a convex course at low relative humidity and a concave course at high relative humidity. To explain this shape of sorption isotherms, various theories have been developed over the past decades (Skaar 1988 pp. 86). Recent studies consider matrix polymer relaxation to explain the sorption behavior of wood (Vrentas and Vrentas 1991; Hill et al. 2012a; 2012b; Popescu and Hill 2013). Following their explanations, the adsorption of water by wood causes a swelling pressure that leads to an expansion of the cell wall nanopores and thereby creates additional sorption sites within the wood matrix. This creation of additional sorption sites is hindered by the limited free volume in the cell wall matrix that restricts the expansion of the nanopores and requires the rearrangement of the matrix polymers. During desorption, the nanopores that are opened by adsorbed water are closed, which requires the relaxation of the surrounding matrix polymers. In this theory, the upward bend in the sorption isotherm of wood at around 60-70 % RH and room temperature is related to the softening of amorphous polymers. Below the glass transition at low relative humidity, the reduced polymer mobility hinders the response of the cell wall matrix to adsorption or desorption of water and delays the opening or closure of the nanopores. Therefore, adsorption and desorption occur in a material that is in different states. When exceeding glass transition by increasing the relative humidity and/or the temperature, the mobility of the cell wall matrix polymers is enhanced. This results in a decrease in hysteresis and in an increase in the additional sorption sites formed upon the expansion of the cell wall nanopores, along with the upward bend in the sorption isotherm. The latter has previously been attributed to sorption of free water in capillaries (Barkas 1937; Kollmann 1962), which was later considered insignificant in wood cell walls below 99.5% relative humidity (Engelund et al. 2010; Thygesen et al. 2010).

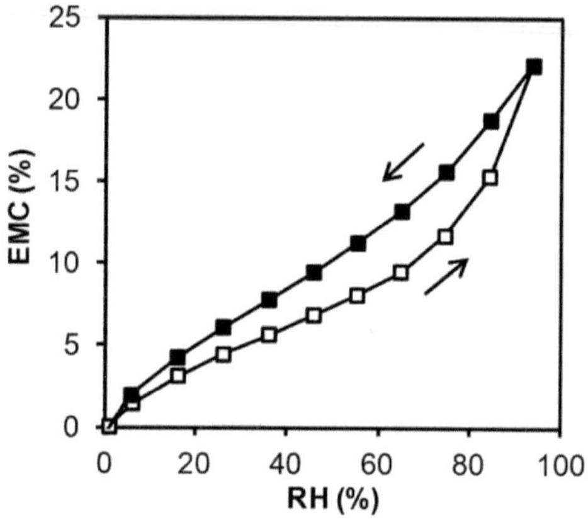

Figure 3: Adsorption (open symbols) and desorption (closed symbols) isotherms of Scots pine sapwood (*Pinus sylvestris* L.) measured at 25°C within the RH range between 0 and 95%.

Below FSP, swelling and shrinkage of wood occurs in response to the uptake and release of water molecules. Swelling and shrinkage of wood is not only depending on the expansion of nanopores within the cell wall matrix, but also on the orientation of the microfibrils within the different cell wall layers. Due to the low MFA in the large S_2 layer, the stiff cellulose microfibrils prevent longitudinal swelling. In a similar way, swelling circumferential to the cell wall is reduced by the high MFA in the S_1 and S_3 layers. Therefore, swelling and shrinkage occur mainly perpendicular to the cell wall (Skaar 1988 pp. 127; Schulgasser and Witztum 2015).

Deterioration of wood by decay fungi

A deterioration of wood can be caused by various organisms, i.e. insects, bacteria and fungi. Fungal decay is very critical, because it can cause a rapid failure of wood products. There is a vast variety of decay fungi that can colonize and degrade wood, including brown rot, white rot and soft rot fungi. The mechanisms involved in the degradation of wood by decay fungi are complex and depend on many factors. In general, decay fungi depolymerize and/or modify cell wall constituents to provide themselves with energy and metabolites for synthesis via metabolism. Major growth requirements for most wood-inhabiting fungi are:

- free water on the surfaces of the cell lumina,
- presence of atmospheric oxygen within the wood,
- a favorable temperature and pH range,
- a digestible substrate that provides energy and metabolites
- chemical growth factors, such as nitrogen compounds, vitamins and essential elements, and
- the absence of toxic extractives (Zabel and Morrell 1992 pp. 90).

In practice, prevention of fungal decay can be achieved by making sure that these requirements are not fulfilled. Fungal decay of wood involves the action of various chemicals, including enzymes. Although the chemicals involved differ considerably between the fungal species, it is a common principle that the wood degrading chemical has to diffuse into the cell wall in order to cause the depolymerization of wood polymers. However, even if a sufficient amount of water as diffusion medium is present, the enzymes involved in fungal degradation are too large to enter the cell wall (Srebotnik et al. 1988; Daniel et al. 1989). Therefore, the fungi must employ smaller diffusible agents to initiate decay, such as reactive oxygen or chelators (Goodell et al. 1997; Hammel et al. 2002).

1.1.3 The need for wood modification

The low dimensional stability and the low biological durability limit the service-life of most European wood species greatly, when applied in exterior conditions in which the requirements for fungal growth are fulfilled. The most common treatment to increase the service-life is the impregnation of these wood species with chemicals that are toxic to microorganisms and/or insects. Inorganic biocides (e.g. based on chromated copper), organic biocides (e.g. quaternary ammonium compounds), or oilborne preservatives (e.g. creosotes) have been used successfully to increase the durability of native wood species for decades (Stirling and Temiz 2014). However, problems arise from some of these chemical treatments such as serious health risks (Willeitner 1973; Jermer and Edlund 1990), leaching of chemicals from treated wood in service (Hingston et al. 2001; Townsend et al. 2005), or difficulties with the disposal of treated wood after service (Voß and Willeitner 1993; Voß and Willeitner 1995). Consequently, there is a tendency of legislation to abandon the use of chemicals that are toxic to humans or the environment, which necessitates the search for alternative treatments.

As an alternative, wood modification provides decay resistance and dimensional stability through non-biocidal modes of action. In contrast to preservative-treated wood, the modified wood should remain non-toxic and no release of toxic components is supposed to take place, neither during service, nor during disposal or recycling (Hill 2006 pp. 20). Wood modification aims at changing the basic chemistry of the wood cell wall polymers to overcome undesired properties of wood, i.e. its low dimensional stability and biological durability (Militz et al. 1997). Treatments with non-toxic chemicals that are deposited in the lumens, but do not diffuse into the cell wall are not in accordance with this definition for wood modification. Although the deposition of chemicals in the lumen can change some wood properties, e.g. the water uptake rate or the hardness (Mundigler and Rettenbacher 2005; Scholz et al. 2009), the chemical and structural composition of the cell wall remains unchanged. Chemical wood modification techniques are based on the impregnation of the wood with small, diffusible chemical agents that can penetrate the cell wall. These chemical agents react with the cell wall polymers and/or themselves, i.e. when exposed to elevated temperatures in a curing step, to become immobilized within the cell wall (Hill 2006 p. 23). The modification intensity for such techniques is typically quantified by the weight of modification agent added in relation to the dry weight of the wood, denoted as weight percent gain (WPG). The improvement in wood properties by such chemical modifications have been explained by different modes of action, such as (1) the blocking of accessible sorption sites to water molecules by hydroxyl substitution (Popper and Bariska 1972; Popper and Bariska 1973; Popper and Bariska 1975), (2) the decrease in the volume of nanopores by cell wall bulking (Stamm 1934; Rowell and Ellis 1978; Hill and Jones 1996), or (3) the formation of cross-links with the cell wall (Tarkow and Stamm 1953; Weaver et al. 1960; Yasuda et al. 1994). Examples for chemical wood modification techniques are acetylation with acetic acid anhydride, furfurylation with furfuryl alcohol, and modification with dimethyloldihydroxyethyleneurea (DMDHEU), which have reached the market under the trade names Accoya®, Kebony® and Belmadur®, respectively.

Thermal wood modification differs from chemical wood modification techniques in that no chemical additives are added to the wood and that the wood weight is reduced by wood degradation, rather than being increased. Thermal modification is thus not limited to wood species that are permeable and easy to impregnate with chemical agents.

1.2 Thermal modification of wood

1.2.1 Thermal modification processes in Europe

It has long been known that properties of wood can be altered by exposure to elevated temperatures, greater than those normally used to dry timber (>120 °C) (Kollmann et al. 1969; Borrega and Kärenlampi 2008a), but considerably lower than those leading to an excessively fast degradation and strong exothermic decomposition of wood (<270 °C) (Stamm et al. 1946). Changes in color and reduction in hygroscopicity (Tiemann 1917), improved dimensional stability (Stamm and Hansen 1937; Seborg et al. 1953), as well as elevated resistance against wood-degrading fungi at the expense of mechanical strength loss (Stamm et al. 1946) of wood after exposure to heat have been known facts for decades. It has also been mentioned in early studies that the presence of oxygen in the treatment reactor carries the risk of excessive strength losses and strong exothermal reactions (Stamm et al. 1946; Sandermann and Augustin 1963a). Today, all thermal modification processes follow the same basic principles that were developed in the first part of the 20th century: The exposure of wood to elevated temperatures (160 – 240 °C) and a minimization of the surrounding oxygen content (Militz and Altgen 2014).

Initial attempts to transfer the scientific knowledge to the development of a commercial thermal modification process were made by Giebeler (1983), but the goal of introducing a large-scale process was ultimately not reached. In the 1990's, however, thermal modification gained new interest when finding alternatives to tropical timbers and preservative-treated wood became increasingly relevant (Militz and Altgen 2014). Until now, various thermal modification technologies have been developed and thermally modified wood is produced in more than 100 production facilities in Europe with a total production volume estimated to more than 300,000 m³ in 2015 (Scheiding 2016). Despite the above described common basic principles of thermal modification, large differences in the applied process conditions and process steps exist between the various technologies that are applied in Europe (Militz and Altgen 2014). Besides peak temperature and duration, the pressure regime and the heat transferring medium are some of the key parameters to distinguish existing thermal modification technologies (**Table 1**).

Table 1: Classification of some industrially applied thermal modification technologies according to the pressure regime and the heat-transferring media applied.

Pressure / Medium	Elevated pressure	Atmospheric pressure	(Partial) vacuum
Heating plates			Vacu³, SmartHeat, ProDeo
Hot oil		Oil-Heat-Treatment (OHT)	
Inert gas	Feuchte-Wärme-Druck (FWD)	Retification	
Superheated steam	WTT, FirmoLin, Bikos-cure	ThermoWood, Stellac, PLATO-cure	WDE-Maspell, Silva-Pro
Saturated steam	PLATO-hydrolysis, Bikos-steam		

The medium in which the thermal modification is conducted not only ensures the transfer of heat to the wood, but usually also reduces the residual oxygen content in the treatment reactor. Although some technologies are based on using hot oil (Sailer et al. 2000; Wang and Cooper 2005; Dubey et al. 2012) or inert gas such as nitrogen (Hofmann et al. 2008; Inari et al. 2009), most thermal modification technologies in Europe apply steam as a medium. However, the process technologies also vary greatly in dependence on the pressure level that is used, ranging from sub-atmospheric (Allegretti et al. 2012; Hofmann et al. 2013) to atmospheric (Viitaniemi et al. 1997; Mayes and Oksanen 2003) and even elevated pressure environments (Giebeler 1983; Willems 2009).

Most industrial thermal modification processes are operated in open reactor systems. One of the first and still most applied process in Europe, the ThermoWood® process, uses superheated steam at atmospheric pressure. In such a process, the relative humidity during the process decreases severely when elevated temperatures are applied. This necessitates a high-temperature (100-130 °C) pre-drying step to reduce the wood moisture content severely, before the actual peak temperature (> 190 °C) is applied. During such a process, a steam flow is continuously channeled through the treatment chamber to remove volatile organic compounds (VOCs) that are emitted from the wood (Mayes and Oksanen 2003). This removal of VOCs is even further facilitated in processes at sub-atmospheric pressure, because the boiling temperature of water and organic compounds is decreased. However, at sub-atmospheric pressure, the transfer of heat to the wood is a critical factor, but can be realized by highly efficient fans for convective heating in low pressure steam (Allegretti et al. 2012) or by conductive heating using heatable plates between the wood layers (Van Acker et al. 2010). Processes in open reactors at (sub-) atmospheric pressure have in common that VOCs are continuously removed, the wood is modified in oven-dry state and that the applied peak temperature usually exceeds 190°C.

Thermal modification processes at elevated pressure in closed reactor systems stand in contrast to the above mentioned processes at (sub-) atmospheric pressure in many ways. Within closed reactor systems, VOCs accumulate and may even accelerate thermal degradation reactions (Stamm 1956), and the pressurized gas atmosphere improves the heat transfer to the wood (Giebeler 1983). A sufficiently high pressure can also prevent rapid drying of wood and if the pressure is generated by water vapor, an equilibrium moisture content ensues as a function of the relative water vapor pressure, even at temperatures exceeding 150 °C (Lenth and Kamke 2001; Kubojima et al. 2003; Ishikawa et al. 2004). In processes at elevated pressure, wood degradation reactions are accelerated in comparison to processes at (sub-) atmospheric pressure conducted at the same temperature level. Thus, the peak temperatures that are applied do usually not exceed 200 °C to avoid excessive degradation. However, differences exist in how the pressure is regulated. During Burmester's FWD (*Feuchte-Wärme-Druck*) process, wood is at a target temperature in an autoclave with a weighed amount of liquid water (Burmester 1973). The FWD process was later further developed by Giebeler (1983) who added an inert gas feed to adjust a constant pressure throughout the process. Today, related processes are operated in Europe, such as the WTT processes with a maximum gauge pressure regulation, or the FirmoLin® process that aims at regulating the relative water vapor pressure by using an external water vapor generator (Willems 2009). Furthermore, there is a process technology (PLATO® process), which combines a process step at saturated water vapor pressure ("hydrothermolysis") with a drying step and a subsequent process step with superheated steam at atmospheric pressure ("curing") (Boonstra 2008).

1.2.2 Macro- and micro-structural defects

In contrast to thermo-hydro-mechanical processing in which wood is shaped, welded or molded through densification, friction or bending whilst wood is softened in the presence of heat (Sandberg et al. 2013), changes to the macro- or microscopic structure of wood are not intended in thermal modification processes. Instead, cracks and other wood defects that arise during the production or the application of wood products are drawbacks that diminish the mechanical strength and increase the water uptake to create optimal moisture conditions for decay fungi (Chang et al. 1982; Eaton and Hale 1993). Cracking thus decreases the market value and the service life greatly.

Cracks are a result of stresses. Wood responds to stresses by creeping that relaxes some of the stresses (Morén 1993). However, if the stresses exceed the fracture strength of wood, cracks are devolved (Lamb 1992). Stresses that result in cracks and other defects in thermally modified wood may be related to the raw material or the conditions during the thermal modification process itself. Raw material based defects, i.e. as a consequence of growth stresses, or defects occurring during the pre-drying in conventional drying chambers, cannot be avoided by optimization of the thermal modification process, but only by careful selection of the material. However, some of these defects, such as cracks developed during early stages of drying, can be very small and not visible to the naked eye, but may grow into visible and more severe cracks at a later stage (Hanhijärvi et al. 2003), e.g. during thermal modification. Wood defects after the thermal modification process are thus not necessarily a result of a flawed modification schedule and it may be difficult to differentiate between defects that arise from the thermal modification process and those that were already present in the raw material.

During thermal modification, stresses may arise from wood drying during the exposure to elevated temperatures. As a consequence of its hygroscopic and anisotropic behavior, wood reacts to changes in moisture content with changes in dimensions and various kinds of deformations. Thermal modification processes in open reactor systems necessitate the drying of the wood to the oven-dry state, at which drying stresses reach a maximum. However, stresses also occur in the absence of wood drying due to the decomposition and modification of wood cell wall constituents at elevated temperatures (Cheng et al. 2007). Without process optimization, severe defects involving cracks, cell collapse and deformations can downgrade the visual wood quality. However, by further optimization of industrial modification processes, especially by avoiding uncontrolled wood drying and by reducing the difference in temperature between the wood surface and the core of boards, the visual wood quality can be reduced to a considerable extent, thus limiting defects to a microscopic scale (Boonstra et al. 2006a; 2006b). Microscopic defects that are reported frequently for thermally modified wood include damaged parenchyma and epithelial cells in rays and resin canals, damages to the pits, cracks running near the compound middle lamella between tracheids and even damaged axial tracheid walls (Boonstra et al. 2006a; 2006b; Sehlstedt-Persson et al. 2006; Awoyemi and Jones 2011; Biziks et al. 2013). Furthermore, *in situ* measurements of Norway spruce wood exposed to temperatures above 200 °C at sub-atmospheric pressure demonstrate a rapid reduction in cell wall thickness (Bernabei and Salvatici 2016). Microscopic defects vary in quantity as well as in type depending on the wood species treated. Broad rays in European beech (*Fagus sylvatica* L.) lead to a high sensitivity towards the formation of radial cracks. Ringporous wood species such as European ash (*Fraxinus excelsior* L.) tend to tangential cracks within the earlywood (Flade 2014). With some softwood species, i.e. Norway spruce (*Picea abies* (L.) Karst.) and Scots pine (*Pinus sylvestris* L.), thermal modification of tangentially sawn boards often results in the delamination (grain raising) between early- and latewood on the growth ring boundary (Boonstra et al. 2006a).

Evaluation of the impact of micro-structural defects on the performance of thermally modified wood is difficult, as it necessitates relating qualitative information on micro-structural aspects to quantitative data on measured wood properties. Nevertheless, some links can be observed. An increase in the capillary water uptake of Scots pine sapwood after mild thermal modification (Johansson et al. 2006; Metsä-Kortelainen et al. 2006) was found to be associated with the damage to the pit membranes in the fenestriform crossfield pits that connect radial ray parenchyma cells with axial tracheids (Sehlstedt-Persson et al. 2006). Furthermore, abrupt fractures of thermally modified spruce and poplar in bending tests were linked to the occurrence of frequent broken tracheid cell walls perpendicular to the fiber direction (Boonstra et al. 2006a; 2006b). However, even though micro-structural features must be considered, their impact on the majority of properties of thermally modified wood is believed to be outweighed by wood chemical changes (Awoyemi and Jones 2011; Militz and Altgen 2014).

1.2.3 Changes in chemical composition and ultra-structure

It is impossible to define a lower temperature limit at which wood chemical changes start, since a large number of additional factors play a major role. The wood moisture content, the oxygen content, the duration of the heat exposure and various other factors affect the chemical reactions and shift the temperature limit required to induce depolymerization of cell wall constituents (Stamm and Hansen 1937; Seborg et al. 1953; Stamm 1956; Sandermann and Augustin 1963b; Mitchell 1988; Borrega and Kärenlampi 2008a). Chemical changes of wood constituents and a loss in substance by thermal degradation can be detected even below 100 °C, if the duration of the heat exposure is sufficiently long (Sandermann and Augustin 1963b). Some chemical changes during natural ageing of wood in the time frame of several decades are very similar to those changes obtained by heating wood at elevated temperatures for several hours (Obataya 2007). However, when considering thermal modification processes during which the duration at a peak temperature is limited to several hours, temperatures above 120 °C are required to obtain significant changes in wood constituents and a measurable loss in wood substance (Kollmann et al. 1969; Borrega and Kärenlampi 2008a).

The qualitative analysis of chemical changes of wood at elevated temperatures is complicated by the broad variety of individual chemical reactions that take place simultaneously within the different wood constituents. However, the chemical reactions within the wood are not only determined by the percentage of individual constituents, but also by their configuration within the cell wall, their chemical bonds or the presence of intermediates derived from other reactions (Sandermann and Augustin 1964; Fengel 1966; Kollmann et al. 1969). Analyses of wood chemical changes during thermal modification processes are further hindered by the high temperatures applied and the corrosive atmosphere caused by the formation of organic acids. The current understanding of the chemical changes during thermal modification processes is thus mainly based on *ex situ* rather than on *in situ* measurements. Nevertheless, an extensive knowledge on wood chemical changes that occur under the conditions applied during thermal modification processes has been gathered within the past decades and is summarized below.

Hemicelluloses

Hemicelluloses are the first cell wall constituents that are affected by chemical changes when increasing the temperature. Their degradation starts with the hydrolytic cleavage of acetyl groups from the side chains, which, in the presence of water, leads to the formation of acetic acid (Sandermann and Augustin 1963b; Carrasco and Roy 1992; Garrote et al. 2001). Acetic acid is thus the most prevalent acid formed upon thermal modification, followed by formic and other organic acids (Sundqvist et al. 2006). The organic acids catalyze the further degradation of the cell wall polysaccharides (Garrote et al. 1999), with hemicelluloses being less thermally stable compared to cellulose due to the lack of crystallinity (Emsley and Stevens 1994). The degradation involves the hydrolysis of the polymer chain into oligomers and monomers as well as the loss in hydrogen bonds (Tjeerdsma et al. 1998; Garrote et al. 1999). The correspondent sugars can be further dehydrated to aldehydes, with furfural and hydroxymethylfurfural (HMF) being typical products originating from pentoses and hexoses, respectively (Fengel and Wegener 1984 pp. 328; Kotilainen 2000; Peters et al. 2009; Karlsson et al. 2012). Furthermore, decarboxylation reactions have a significant effect on carbohydrate degradation and lead to the formation of carbon dioxide (Demirbaş 2000; Willems et al. 2013). However, the structure of hemicelluloses is heterogeneous with differences in the thermal behavior of the individual hemicellulosic constituents. In general, pentoses (e.g. xylan) are less temperature stable than hexoses (e.g. galactoglucomannan) (Kotilainen 2000).

Cellulose

Cellulose is not affected as strongly by thermal degradation as hemicelluloses, although many chemical reactions are similar (Sandermann and Augustin 1963a; Fengel 1967; Alén et al. 2002). The highly ordered, crystalline regions are the most thermally stable and hydrolysis-resistant parts of the cellulose microfibrils (Fengel 1967). The hydrolytic cleavage starts in the less ordered, amorphous regions of the cellulose microfibrils, which results in an increase in the relative crystallinity of the cellulose on the one hand (Sivonen et al. 2002; Wikberg and Maunu 2004), and in a decrease in the DP on the other hand (Fengel 1967). The removal of amorphous cell wall polymers decreases the order in the arrangement of the cellulose microfibrils and results in an increased cellulose aggregation, while the MFA of the cellulose microfibrils remains unchanged (Andersson et al. 2005; Salmén 2015).

Lignin

When thermal degradation is determined gravimetrically, lignin is the most thermally stable cell wall component, leading to an increase in the relative percentage of lignin in the wood after thermal modification (Sandermann and Augustin 1963a; Kollmann and Fengel 1965; Alén et al. 2002). Nevertheless, chemical changes occur in the lignin as well. Thermal degradation of lignin in the presence of heat and moisture has been shown to involve a competition between depolymerization and repolymerization reactions (Li et al. 2007). Thermal modification causes a cleavage of the methoxyl groups and a depolymerization of the lignin macromolecule to low molecular weight compounds, i.e. by cleavage of ether linkages (Tjeerdsma et al. 1998; Sivonen et al. 2002; Nuopponen 2005). Simultaneously, condensation reactions involving not only lignin, but also furfural and other degradation products take place (Tjeerdsma et al. 1998; Garrote et al. 1999). Furthermore, stable free radicals are formed within the lignin during the process, which are believed to have a semiquinonic structure (Sivonen et al. 2002; Ahajji et al. 2009; Willems et al. 2010).

Extractives

The majority of extractives found in unmodified wood is degraded or becomes volatile during the exposure to elevated temperatures. For Scots pine thermally modified in an open reactor system, the movement of fats and waxes along the axial parenchyma cells to the surface was detected in the temperature range between 100 and 160 °C and they disappeared from the surface above 180 °C. Above 200 °C, even the resin acids were found to disappear (Nuopponen et al. 2003). It should be noted, however, that while many native extractives are removed by thermal modification, thermal degradation products may partly remain within the wood as new extractives (Esteves et al. 2008; Poncsak et al. 2009; Esteves et al. 2011).

1.2.4 Mass loss by thermal modification

The degree of thermal degradation is often determined gravimetrically based on the change in dry mass of the wood during the process. In cases where oven-drying of the actual test material needs to be avoided, the dry mass is calculated rather than being measured directly. This calculation is based on the mass of the material before and after the process and the respective moisture content, which is either measured for a separate sample set or for samples taken from the test material before and after the process.

Mass loss during the process is often used as a marker for the impact of process conditions on thermal degradation and is strongly correlated with the change in properties during the thermal modification process (Welzbacher et al. 2007; Chaouch et al. 2013). This quantification of thermal degradation is based on the depolymerization of cell wall constituents, i.e. hemicelluloses, and their formation into degradation products, such as furfural, HMF, organic acids or carbon dioxide, which vaporize at elevated temperatures to result in a loss in wood substance (Fengel 1966; Alén et al. 2002). However, this quantification of thermal degradation by wood mass loss suffers from inaccuracies. The majority of native extractives are removed from the wood at comparable low temperatures (< 200 °C) when performing the thermal modification process at (sub-) atmospheric pressure (Nuopponen et al. 2003; Poncsak et al. 2009). This causes a loss in dry wood mass, even if no degradation of cell wall constituents takes place. As an example for such an effect, the thermal modification of Scots pine heartwood results in a higher mass loss compared to Scots pine sapwood, because the heartwood contains a larger amount of extractives that are emitted at elevated temperatures (Metsä-Kortelainen 2011). Furthermore, the mass loss does not fully account for thermal degradation products that remain within the wood, or for chemical reactions, such as condensation or esterification reactions, that do not result in a strong loss in mass but in a fixation of some degradation products (Garrote et al. 1999; Tjeerdsma and Militz 2005).

1.2.5 Water-related properties

Wood water relations are strongly influenced by thermal modification. A reduction in hygroscopicity of wood after the exposure to elevated temperatures was already discovered by Tiemann (1917) and has since then been further evidenced in a vast number of studies (see e.g. Seborg et al. 1953; Tjeerdsma et al. 2002; Popper et al. 2005; Jalaludin et al. 2010; Hill et al. 2012b; Olek et al. 2013). The equilibrium moisture content (EMC), which is the ratio of the mass of moisture within a sample that is conditioned at a specific climate to the dry mass of the sample, is often used as measure for the hygroscopicity. By thermal modification of wood, the EMC decreases in dependence on the process conditions applied, i.e. with increasing peak temperature and duration. The reduction in EMC is thus strongly correlated with the mass loss during the modification process. As a function of mass loss, the EMC of wood thermally modified at atmospheric pressure in superheated water vapor typically decreases almost linearly at a low mass loss range, but reaches saturation at a higher mass loss range (Viitaniemi et al. 1997; Esteves et al. 2007).

Since the wood water relations are still not fully understood (Engelund et al. 2013), explaining the mechanisms that cause the reduction in hygroscopicity during the thermal modification process is still a subject for research. The reduced hygroscopicity is often explained by a decrease in the water-accessible OH groups as a result of the chemical changes during the process. Cell wall polysaccharides have the largest share on the OH groups present within the wood (Runkel 1954; Runkel and Lüthgens 1956). The removal of cell wall polysaccharides by thermal degradation should thus result in a reduction in the total amount of OH groups of wood. As the removed cell wall polysaccharides are preferentially those that have the highest mobility and reactivity (Borrega and Kärenlampi 2008a), water-accessible OH groups are predominantly eliminated (Willems 2014a). It has further been suggested that the accessibility of the remaining OH groups is additionally reduced by the increase in cross-linking in the lignin-carbohydrate-complex. This enhanced cross-linking limits the expansion of cellulose microfibrils and thereby reduces the capacity to adsorb water between the cellulose chains (Tjeerdsma et al. 1998).

However, using the hydrogen-deuterium-exchange method, Rautkari et al. (2013) found poor correlations between the accessible OH group content and the EMC of thermally modified wood. This leads to the conclusion that further mechanisms must apply. The formation of irreversible hydrogen bonds and a more tightly bonded wood structure as a result of wetting and re-drying at elevated temperatures, denoted as hornification (Borrega and Kärenlampi 2010), might be such an additional mechanism. An alternative mechanism, based on the consideration of sorption processes in a glassy polymer matrix, was suggested by Hill et al. (2012b). This mechanism postulates that the EMC of wood is determined by the cell wall modulus. The adsorption of water by wood creates a swelling pressure that causes the expansion of the cell wall nanopores and thereby enables the accessibility of further OH groups. This expansion of cell wall nanopores is resisted by the limited void volume and the stiffness of the cell wall matrix (see also **1.1.2**). The EMC can thus be interpreted as the balance between the swelling pressure which is depending on the surrounding water vapor on the one hand, and the internal restraint of the cell wall which is depending on the cell wall modulus, on the other hand. Increasing the cell wall modulus and the stiffness of the wood cell wall by thermal modification would consequently restrict the expansion of nanopores and thereby explain the reduction in EMC.

Although several attempts to explain the sorption properties of thermally modified wood exist, the mechanisms involved in some sorption phenomena have not been fully understood, yet. Using a dynamic vapor sorption (DVS) apparatus, Hill et al. (2012b) found that the sorption isotherms changed between the first and subsequent sorption cycles. During the first sorption cycle, the sorption hysteresis (difference between adsorption and desorption isotherms) was higher for thermally modified wood than for unmodified wood, but decreased to the values determined for unmodified wood in the second sorption cycle. Furthermore, Rautkari and Hill (2014) showed that higher initial wood MCs of Scots pine sapwood during thermal modification in saturated water vapor increased the mass loss generated during the process, but resulted in a lower reduction in EMC. To date, there is still a lack of conclusive explanations for these two phenomena.

Altered wood-water relations by thermal modification are also expressed by a change in the wettability. The wettability of wood by water, assessed by contact angle measurements, decreases by thermal modification in dependence on the process conditions applied (Pétrissans et al. 2003; Hakkou et al. 2005; Gérardin et al. 2007). Free reactive OH groups in unmodified wood may be strongly involved during the wetting of wood by water, since they are likely to form hydrogen bonds with water molecules. Reducing the amount of OH groups by the thermal degradation of hemicelluloses is thus a likely cause for the decrease in wettability of the wood by water during thermal modification (Gérardin et al. 2007). However, Hakkou et al. (2005) found a severe decrease in wettability by water after a thermal modification of wood in the temperature range between 100 and 160 °C although no mass loss was detected. Therefore, they concluded that the thermal degradation of hemicelluloses cannot be the sole reason for the decrease in wettability and that further mechanisms, such as the plasticization and re-organization of wood polymers during the modification process, need to be considered.

The decrease in wettability also reduces the capillary forces during liquid water uptake. Consequently, the rate at which the wood takes up liquid water decreases with increasing peak temperature and duration applied during the thermal modification process (Johansson et al. 2006; Metsä-Kortelainen et al. 2006). However, as mentioned above (see **1.2.2**), micro-structural defects in some wood species might compensate for the decrease in capillary forces to some extent by increasing the number of capillaries and enhancing the penetration of water from cell to cell (Sehlstedt-Persson et al. 2006). As a an example, thermal modification of Scots pine sapwood at mild temperatures even enhances the capillary water uptake, and comparable high mass loss levels at processes well above 200 °C are required to reduce the uptake of water (Johansson et al. 2006; Metsä-Kortelainen et al. 2006).

1.2.6 Dimensional changes

Although the thermal modification of wood causes a degradation of cell wall constituents (see **1.2.3**), the porosity of the wood in dry state does not necessarily increase and may even decrease after the modification process (Zauer et al. 2013). Voids created by the removal of cell wall components close during the thermal modification and result in a reduction in the dry dimensions (after oven-drying above 100 °C) of wood (Burmester 1975; González-Peña et al. 2009). The annealing and rearrangement of amorphous cell wall polymers at elevated temperatures might be responsible for this closure of voids (Obataya and Tomita 2002). Tangential shrinkage stresses have indeed been detected in wood treated in high-temperature, saturated water vapor environments, in which no drying, but degradation of cell wall polymers takes place (Cheng et al. 2007). González-Peña et al. (2009) suggested that by preferential removal of side chain sugars in hemicelluloses, the residue occupies less space within the cell wall, and that a similar effect is achieved with the more crystalline cellulose and the more condensed lignin. Interestingly, they observed a higher reduction in dimensions in the tangential direction than in the radial direction. They hypothesized that radial cell walls contain more lignin than tangential cell walls and are thus less susceptible for thermal degradation and mass loss. In contrast, the dimensions in longitudinal direction did not decrease, but even increased slightly for mass losses below 10 %. The cited authors assumed that this increase in longitudinal dimensions is linked to stresses caused by the dimensional changes in the transverse direction.

By measuring the wet dimensions (at or above fiber saturation) in addition to the dry dimensions, swelling can be determined as follows:

$$\text{Swelling } (\%) = \frac{(\text{Wet dimensions} - \text{Dry dimensions})}{\text{Dry dimensions}} \times 100 \qquad \text{(Eq. 1)}$$

In relation to the original dimensions of wood before thermal modification, the reduction in wet dimensions is even higher than the reduction in dry dimensions. Consequently, the swelling is reduced, which makes thermally modified wood more dimensional stable than unmodified wood (Burmester 1975). There is a correlation between the dimensional stability and the mass loss generated by the modification process. For thermal modification processes of wood in dry state, the dimensional stability, quantified by the anti-swelling efficiency (ASE), increases strongly at a low mass loss range, but reaches saturation at a higher mass loss range (Stamm et al. 1946; Esteves et al. 2007), comparable to the EMC dependence on mass loss (see **1.2.5**). However, although indicated by the mass loss correlation, the preferential removal of hemicelluloses as the most hygroscopic cell wall component is not the main factor for the improvement in dimensional stability (Repellin and Guyonnet 2005). The removal of hemicelluloses by treatments of wood in sulfuric acid even increases the swelling of wood upon water-soaking, because voids created by the removal of hemicelluloses are closed for wood in dry state, but are occupied by water in wet state (Burmester 1975). In a similar way, increasing the wood MC during the thermal modification process increases the wood mass loss, but hinders the improvement in dimensional stabilization during the thermal modification process (Stamm and Hansen 1937; Seborg et al. 1953; Rautkari and Hill 2014). From this it follows that there must be an alternative mechanism other than the loss in cell wall components that leads to a dimensional stabilization of wood by thermal modification. Ultra-structural changes and modifications within the lignin carbohydrate complex have been suggested as potential mechanisms (Repellin and Guyonnet 2005), but require further verification and investigation. Furthermore, it was recently shown that thermal degradation products that remain within the cell wall can cause a cell wall bulking effect. The degradation products occupy nanopores that can thus not be fully filled by water and thereby contribute to an enhanced dimensional stability. Consequently, the dimensional stability of thermally modified wood decreases upon the extraction of soluble compounds (Biziks et al. 2015; Čermák et al. 2015).

1.2.7 Mechanical properties

In practical applications, material inherent features, such as the density, growth characteristics or other features, have a strong influence on the mechanical properties of wooden products. However, when studying the effect of process conditions during the thermal modification on wood mechanical properties, such variability needs to be reduced by applying standardized test methods on small, clear and defect-free specimens of wood. By thermal modification, many mechanical properties are reduced, although the extent of this property change differs. With regard to the three primary stresses, tensile strength is usually affected the most by thermal modification, followed by shear and compressive strength (Boonstra et al. 2007a; Korkut et al. 2008a; Korkut et al. 2008b). In longitudinal direction, the compressive strength may even increase by thermal modification (Boonstra et al. 2007a). Since it is a combination of the three primary stresses, the bending strength is reduced as well. It can decrease to half of its original value for mass losses considerable below 20 % (Esteves et al. 2007; Borrega and Kärenlampi 2008b). Interestingly, mild thermal modification processes that result in minor mass losses may even increase the bending strength (Kubojima et al. 2000; Rautkari et al. 2014). Since mechanical tests are usually applied on specimens that are conditioned at a specific climate, the reduced hygroscopicity of thermally modified wood might compensate for its reduction in strength to some extent (Borrega and Kärenlampi 2008b; Arnold 2010). It has also been suggested that the increase in compressive strength outweighs the effect of the decrease in tensile strength during a bending test of wood modified in a mild thermal modification process (Boonstra et al. 2007a).

In contrast to wood strength, the effect of thermal modification on the modulus of elasticity (MOE) is limited, with only slight reductions even at comparable high peak temperatures or long peak durations (Kubojima et al. 2000; Boonstra et al. 2007a). Similar to the bending strength, MOE of thermally modified wood can be even higher than the original value, indicating an increase in stiffness (Kubojima et al. 2000; Borrega and Kärenlampi 2008b; Rautkari et al. 2014).

A drastic reduction by the exposure to elevated temperatures is, however, evident for the work to maximum load in bending (LeVan et al. 1990; Kubojima et al. 2000; Borrega and Kärenlampi 2008b). Since the work to maximum load is a function of both, stress and strain, it is usually affected long before strength properties or properties related to the proportional limit, such as MOE, are significantly affected (Winandy and Rowell 2005). In a similar way, other measures for the toughness of wood, such as the impact bending strength, are strongly affected by thermal modification as well (Kubojima et al. 2000; Boonstra et al. 2007a; Wetzig et al. 2012). Remarkably, the reduction in toughness by thermal modification is considerably higher in the inelastic region above the proportional limit than in the elastic region of the load-deflection curve (Kubojima et al. 2000; Phuong et al. 2007; Borrega and Kärenlampi 2008b). This evidences that the wood changes from a viscoelastic and ductile to a more brittle and glassy material as a consequence of the thermal modification (Kubojima et al. 2000; González-Peña and Hale 2007; Phuong et al. 2007).

One simple reason for the decrease in strength and toughness might be the loss in substance that occurs during thermal modification (see **1.2.4**). Stresses applied by an external load are thereby distributed over less cell wall material (Boonstra et al. 2007a). However, the loss in bending strength and toughness is considerably higher than the respective mass loss during the thermal modification (Esteves et al. 2007; Borrega and Kärenlampi 2008b). Therefore, the reasons for the considerable changes in the mechanical properties are more likely to be found in the effects of thermal modification processes on the chemical and (ultra-) structural composition of the wood cell wall (Boonstra et al. 2007a; Arnold 2010; Hughes et al. 2015). Although cellulose is considered as the reinforcement in wood as a biopolymer composite, the incipient loss in strength (up to 30-40% loss in bending strength) upon exposure to elevated temperatures cannot be linked to the degradation of cellulose or its decrease in DP, but rather to the degradation of hemicelluloses. It was suggested that the interpolymeric load sharing capability of the wood cell wall composite is reduced as side-chain hemicelluloses constituents (e.g. arabinose and galactose) and thereafter as the main-chain hemicellulose constituents (e.g. xylose and mannose) are degraded (LeVan et al. 1990; Sweet and Winandy 1999; Winandy and Lebow 2001). However, even though significant thermal degradation of cellulose and lignin start at a later stage of wood strength loss, rearrangement and reorientation of the molecular structure of cellulose and lignin might contribute to incipient changes in mechanical properties. Enhanced cross-linking of the lignin network might enhance the stiffness of the cell wall matrix, which would increase the MOE. Furthermore, the percentage of crystalline cellulose increases as a consequence of the preferential degradation of amorphous polysaccharides. Thereby, the cell wall might become more rigid, causing the wood to break more easily by tension, but enhancing its compressive strength in longitudinal direction (Boonstra et al. 2007a).

Besides chemical reactions, physical changes within the wood cell wall might affect the mechanical properties of thermally modified wood additionally. Ultra-structural rearrangements between wood polymers as a consequence of drying were suggested to increase the stiffness of the wood cell wall matrix (Suchy et al. 2010b). Furthermore, Borrega and Kärenlampi (2008b) performed a bending test on wood that was thermally modified under variation of the relative humidity and found that samples modified at an intermediate relative humidity (50-61%) featured a higher bending strength and inelastic toughness than samples that were thermally modified at low relative humidity (< 15%) or saturated water vapor. They concluded that irreversible hydrogen bonding as a result of wetting and re-drying at intermediate relative humidity has a beneficial impact on the mechanical properties.

1.2.8 Biological durability

There is a strong impact of the thermal modification process on the resistance of wood against decay fungi. The wood mass loss caused by fungal decay during the incubation with wood degrading basidiomycetes is drastically reduced for thermally modified wood. This reduction is depending on the peak temperature and duration applied and the mass loss generated by the thermal modification (Hakkou et al. 2006; Boonstra et al. 2007c; Šušteršic et al. 2010; Mohareb et al. 2012). Achieving a very durable wood material (durability class 1, EN 350-1: 1994) from non-durable wood species is possible, but the process conditions (e.g. peak temperatures and durations) that are required for such a large improvement in biological durability are usually accompanied by a strong reduction in mechanical strength and ductility (Welzbacher et al. 2007; Andersons et al. 2010). Nevertheless, the resistance against basidiomycetes of thermally modified wood is still somewhat affected by the raw material. As an example, the increased biological durability of Scots pine heartwood compared to Scots pine sapwood is still evident after the thermal modification process (Boonstra et al. 2007c; Metsä-Kortelainen and Viitanen 2009).

In laboratory monoculture tests on agar-malt medium (e.g. EN 113: 1996; CEN/TS 15083-1: 2005), *Rhodonia placenta* has often been identified as the most aggressive decay fungi against thermally modified wood (Welzbacher and Rapp 2007; Boonstra et al. 2007c). In contrast to laboratory tests with basidiomycetes on an agar-malt medium, thermally modified wood does often not increase the resistance against decay fungi during laboratory tests in ground contact sufficiently (Kamdem et al. 2002). Results on the biological durability from laboratory tests are confirmed by different field tests. Field tests aiming at a European hazard class 3 (EN 335-1: 2006), e.g. the double layer test and the lap joint test, prove the enhanced resistance against decay fungi in above ground conditions (Welzbacher and Rapp 2007; Metsä-Kortelainen et al. 2011). Field tests according to EN 252: 2014, which aim at a European hazard class 4, validate the insufficient decay resistance of thermally modified wood in ground contact (Welzbacher and Rapp 2007).

The mode of action of the enhanced resistance against decay fungi of thermally modified wood in above-ground conditions is still a subject for research. Based on monoculture tests with *R. placenta*, it was shown that the degradation of thermally modified wood is inhibited initially and that once that the fungal degradation starts, the rate is lower than in unmodified wood. However, the fungus uses the same mechanism to degrade thermally modified wood as for degrading unmodified wood (Ringman et al. 2016). There is no evidence that a formation of toxic compounds during the modification process contributes to an improved resistance against basidiomycetes (Kamdem et al. 2000; Kamdem et al. 2002; Hakkou et al. 2006; Mohareb et al. 2012). Instead, the enhanced durability of thermally modified wood is explained by changes in the chemical composition of the wood cell wall during the thermal modification process. The most prominent reasons that have been suggested are: (a) the removal of hemicelluloses as a main nutrient source for decay fungi (Hakkou et al. 2006); (b) the modification of the wood polymers, which hinders their recognition by enzymes involved in fungal degradation (Weiland and Guyonnet 2003; Hakkou et al. 2006); (c) the inhibited penetrability of the cell wall for diffusible agents emitted by the fungi in the initial stage of decay caused by the reduced fiber saturation and reduced swelling of the modified wood (Boonstra et al. 2007c; Thybring 2013; Zelinka et al. 2014; Hosseinpourpia and Mai 2016; Zelinka et al. 2016); and (d) the presence of stable free radicals in thermally modified wood, which may act as antioxidants and quench reactive oxygen species emitted by the fungi at the initial stage of decay (Willems et al. 2010).

1.2.9 Weathering and coating performance

Besides biotic factors such as fungal decay, the service life of wooden products is also strongly determined by abiotic factors involved in weathering. During weathering, the reduced water uptake (see **1.2.5**) of thermally modified wood should reduce the stresses caused by the exposure to rain, humidity and sunlight. However, surface cracking during outdoor exposure is typically not improved by thermal modification. Studies report on similar cracking (Jämsä et al. 2000), or even on increased cracking and grain-raising (Feist and Sell 1987) of thermally modified Norway spruce and Scots pine compared to unmodified control boards during natural weathering tests. There are also reports from material in service with exceptional frequent defects for thermally modified wood products, such as grain-raising of softwood species, as well as radial surface or internal cracks in hardwood species (Flade 2014).

Although thermal modification does not explicitly aim at improving the resistance of wood against photodegradation, an improvement would be advantageous with respect to maintaining its characteristic brown color in exterior applications. In unmodified wood, lignin is the main UV light absorbing cell wall component and is thus strongly affected by photodegradation (Norrström 1969). Since the percentage of lignin as the most susceptible wood component increases on the one hand, but is also modified by various chemical reactions on the other hand, it is difficult to anticipate whether or not the resistance against photodegradation is affected by the thermal modification process. For short durations of artificial weathering of 72 h (Huang et al. 2012) or up to 835 h (Ayadi et al. 2003), a higher resistance of thermally modified wood was found and assigned to the chemical alteration of the lignin and the formation of phenolic compounds. It was also observed that photodegradation products of thermally modified wood do not leach out as easily as for unmodified wood (Nuopponen et al. 2004). However, during long-term natural or artificial weathering, the lignin content diminishes almost completely and the surface color is not stable and turns grey, similar to the behavior of unmodified wood (Jämsä et al. 2000; Huang et al. 2012; Yildiz et al. 2013).

To prevent photodegradation and cracking, a surface treatment of thermally modified wood with coating systems can be required. However, studies on the performance of coating systems on thermally modified wood surfaces have varying results, which might be related to the different modification technologies, wood species or coating systems used. There are studies suggesting that no alterations in the coating recommendations are required and that thermally modified wood as a substrate does not differ from unmodified wood (Jämsä et al. 2000; Grüll et al. 2010). The service life of exterior coatings should even be prolonged, since thermally modified wood reduces stresses within the coating that are caused by the movement of the substrate related to swelling and shrinkage (Podgorski and Roux 1999). A further advantage might be the removal of native wood extractives by thermal modification (Poncsak et al. 2009), which can cause discoloration of coatings and surface deactivation during ageing (Nussbaum 1999; Nussbaum 2004). However, there are also studies reporting on poor performances of selected coating systems on thermally modified wood as a substrate (Feist and Sell 1987; de Moura et al. 2013). The weakening of the substrate due to the loss in strength (De Moura et al. 2013) as well as the increased hydrophobicity (Gérardin et al. 2007; Metsä-Kortelainen and Viitanen 2012) are suggested as potential reasons for such poor performances.

1.3 Scope of this thesis

The aim of this thesis was to evaluate the impact of different process conditions of thermal modification processes on the properties and the performance of the resulting wood material.

The commercial implementation of thermal wood modification processes already started in the early 1990's, and up to today thermally modified wood has been studied extensively, as reviewed in the previous chapter (see **1.2**). Nevertheless, the changes in wood properties and performance, the underlying modes of action and their link to specific process conditions of the various techniques to thermally modify wood are still not fully understood. This can be partly explained by the lack of long-term experience with thermally modified wood in practical conditions, which now, after many years of commercial application, has lead to new questions to be answered. In addition to long-term experiences with thermal modification technologies that have existed for many years, further questions arise from the development of new technologies that differ in the process conditions applied. In particular, closed reactor systems have been developed that enable the use of elevated water vapor pressure in a one step thermal modification process (Willems 2009). However, up to today, there is still a lack of scientific research on the impact of elevated water vapor pressure during the modification process on the obtained product and it remains a question if the modes of action and the resulting properties differ from wood treated in open reactor systems.

The investigations performed in the framework of this thesis are divided into two parts:

The first part of this thesis (**publications I-III**) investigates the thermal modification of wood in closed reactor systems. It focuses on the impact of elevated water vapor pressure on the thermal wood degradation and the underlying chemical reactions as well as on reversible and irreversible changes in the physical wood properties. A special emphasis is on the change in wood sorption by thermal modification. The application of elevated water vapor pressure adds another process parameter that allows for the investigation of effects that cannot be assessed by the simple variation of peak temperature and duration during thermal modification processes, i.e. the effect of elevated wood moisture contents and/or the effect of wood drying.

The second part of this thesis (**publications IV-VI**) provides understanding why the performance of thermally modified wood in exterior applications cannot entirely meet the expectations derived from its enhanced resistance against decay fungi. Problems in exterior applications might arise from the impact of abiotic factors during the weathering of the wood surface or changes in the coatability of wood by thermal modification. To assess these issues, material coming from the industrial-scale production in open reactor systems was investigated with regard to: (1) the influence of different steps of the production as well as capillary wetting cycles on the development of surface cracks, (2) the impact of different modification temperatures on the photodegradation of the wood surface, and (3) material characteristics that may influence the performance of coating systems.

1.4   List of publications in peer-reviewed journals

The thesis is based on the following publications in peer-reviewed journals.

Publication I

> **Wood degradation affected by process conditions during thermal modification of European beech in a high-pressure reactor system**
> Altgen M., Willems W., Militz H.; European Journal of Wood and Wood Products, (2016), Vol. 74(5):653-662

Publication II

> **Influence of process conditions on hygroscopicity and mechanical properties of European beech thermally modified in a high-pressure reactor system**
> Altgen M., Militz H.; Holzforschung, (2016), Vol. 70(10):971-979

Publication III

> **Wood moisture content during thermal modification process affects the improvement in hygroscopicity of Scots pine sapwood**
> Altgen M., Hofmann T., Militz H.; Wood Science and Technology, (2016), Vol. 50(6):1181-1195

Publication IV

> **Wood defects during industrial-scale production of thermally-modified Norway spruce and Scots pine**
> Altgen M., Adamopoulos S., Militz H.; Wood Material Science & Engineering, *in press*

Publication V

> **Photodegradation of thermally-modified Scots pine and Norway spruce investigated on thin micro-veneers**
> Altgen M., Militz H.; European Journal of Wood and Wood Products, (2016), Vol 74(2):185-190.

Publication VI

> **Thermally-modified Scots pine and Norway spruce wood as substrate for coating systems**
> Altgen M., Militz H.; Journal of Coatings Research and Technology, *accepted*

Author's contribution

Michael Altgen designed the experimental setup in all the papers (in cooperation with Wim Willems (I) and Stergios Adamopoulos (IV)). He was responsible for conducting the experiments (in cooperation with Tamás Hofmann (III) Stergios Adamopoulos (IV)). Michael Altgen is the main-author in all of the publications, with supervision by Holger Militz.

1.5 Other publications on thermally modified wood

Bestimmung der Behandlungsqualität von Thermoholz mithilfe von Schnellverfahren. Teil 1: Elektronenspin-Resonanz-Spektroskopie
Altgen M., Welzbacher C., Humar M., Willems W., Militz H.; Holztechnologie, (2012), 53(6):44-49

Factors influencing the crack formation in thermally modified wood
Altgen M., Adamopoulos S., Ala-Viikari J., Hukka A., Tetri T., Militz H., Proceedings of the 6th European Conference on Wood Modification, (2012), Ljubljana, Slovenia

Bestimmung der Behandlungsqualität von Thermoholz mithilfe von Schnellverfahren. Teil 2: Nah-Infrarot-Spektroskopie
Altgen M., Welzbacher C., Militz H.; Holztechnologie, (2013), 54(1):40-44

Surface performance of thermally modified wood during weathering
Altgen M., Ala-Viikari J., Hukka A., Tetri T., Militz H., Proceedings of the joint conference of COST FP0904 and FP1006: "Characterization of modified wood in relation to wood bonding and coating performance", (2013), Rogla, Slovenia

Processes and properties of thermally modified wood manufactured in Europe
Militz H., Altgen M.; In: Schultz T.P., Goodell B., Nicholas D.D. (Ed.) Deterioration and protection of sustainable biomaterials; ACS Symposium Series 1158, (2014), American Chemical Society, Washington, DC

Performance of water-borne coating systems on thermally modified wood
Altgen M., Ala-Viikari J., Hukka A., Tetri T., Militz H., Proceedings of the 7th European Conference on Wood Modification, (2014), Lisbon, Portugal

Comparison of EMC and durability of heat treated wood from high versus low water vapour pressure reactor systems
Willems W., Altgen M., Militz H.; International Wood Products Journal, (2015), 6(1):21-26

Effect of temperature and steam pressure during the thermal modification process
Altgen M., Militz H., Proceedings of the 8th European Conference on Wood Modification, (2015), Helsinki, Finland

Quality control methods for thermally modified wood
Willems W., Lykidis C., Altgen M., Clauder L.; Holzforschung, (2015), 69(7):875-884

Chapter 2   Publication I

Wood degradation affected by process conditions during thermal modification of European beech in a high-pressure reactor system

Altgen M.[1]*, Willems, W.[2], Militz H.[1]

[1]*Wood Biology and Wood Products, Burckhardt Institute, Georg-August University Göttingen, Büsgenweg 4, 37077 Göttingen, Germany*

[2]*FirmoLin Technologies BV, Grote Bottel 7a, 5753 PE Deurne, Netherlands*

corresponding author:
Phone: +49-551 3933664
Fax: +49 551 399646
E-mail: maltgen@gwdg.de

Originally published in:
European Journal of Wood and Wood Products
Springer-Verlag Berlin Heidelberg
ISSN (Online) 1436-736X
ISSN (Print)   0018-3768
DOI: 10.1007/s00107-016-1045-y

Received: 11 August 2015 / Accepted: 07 April 2016 / Published online: 15 April 2016

Abstract: The degradation of beech wood during a thermal modification process in a high-pressure reactor system using steam as medium was investigated. The wood was modified at different peak temperatures (150-180 °C), peak durations (1-6 h) and maximum water vapor pressures (0.14-0.79 MPa), while wood mass loss and wood moisture content as well as soluble degradation products were analyzed. Wood degradation was found to be predominantly determined by the maximum pressure, rather than the peak temperature applied. However, accumulation of degradation products, i.e. carbohydrates and furfural, in wood modified at elevated pressure had to be considered when using mass loss as a marker for wood degradation. Mass loss and mass loss rate increased with the maximum pressure until reaching saturation at mass losses above 20 %, due to the limited amount of amorphous carbohydrates within the wood. Several factors have been discussed with regard to their impact on accelerated degradation reactions at elevated water vapor pressure, such as a better heat transfer in a compressed gas atmosphere, reduced evaporative cooling, the accumulation of organic acids as well as the presence of water in the wood during the process. However, none of these individual factors were completely consistent with the observed mass loss progression, which leads to the conclusion that the impact of elevated water vapor pressure, rather, is a combination of several factors that apply simultaneously. The application of elevated pressure might enable an effective process technique to generate sufficient wood degradation to upgrade dimensional stability and biological durability of wood at a low temperature range.

Keywords: closed reactor system, mass loss, thermally-modified wood, water vapor pressure

2.1 Introduction

The effect of exposing wood to elevated temperatures to alter its properties has been extensively studied over the past few decades and has led to a number of commercialized thermal modification processes, as reviewed by Hill (2006), Esteves and Pereira (2009) and Militz and Altgen (2014). Thermal modification processes result in degradation of wood (loss of substance) in response to high temperature treatment. The thermal degradation of wood reduces mechanical strength and ductility (Kubojima et al. 2000), but also improves dimensional stability (Popper et al. 2005) and resistance to decay fungi (Kamdem et al. 2002).

Wood degradation already starts at low temperatures with cleavage of the acetyl groups of hemicelluloses and the formation of carboxylic acids, such as acetic and formic acid (Garrote et al. 2001; Tjeerdsma and Militz 2005; Sundqvist et al. 2006). These acids catalyze the further degradation of amorphous carbohydrates at elevated temperatures, i.e. by hydrolysis and dehydration, which result in the formation of aldehydes, with furfural and hydroxymethylfurfural being the main degradation products of pentoses and hexoses, respectively (Tjeerdsma et al. 1998; Kotilainen 2000). Besides hydrolysis and dehydration, decarboxylation reactions have a significant effect on carbohydrate degradation and lead to the formation of carbon dioxide which is released from the wood (Demirbaş 2000; Willems et al. 2013). In contrast to the carbohydrates, lignin is more resistant to thermal degradation, but chemical changes such as the cleavage of ether linkages and demethoxylation on the one hand, and cross-linking and further condensation reactions on the other hand, take place (Tjeerdsma et al. 1998; Sivonen et al. 2002; Wikberg and Maunu 2004; Nuopponen 2005).

Process conditions, such as the peak temperature and duration, the wood moisture content or the heat transfer media can severely affect wood degradation and consequently property changes (Mitchell 1988; Alén et al. 2002; Welzbacher et al. 2007; Candelier et al. 2013). Understanding the impact of process conditions is therefore essential for the production of thermally-modified wood that fulfills the requirements relevant for a given end-use application. Among the various process conditions, the pressure regime might affect the thermal modification process considerably. Commercially applied thermal modification processes operate at sub-atmospheric (e.g. SmartHeat®, Vacu³®), atmospheric (e.g. ThermoWood®) or elevated pressure regimes (e.g. FirmoLin®, WTT). During processes under (sub-) atmospheric pressure regimes, wood degradation products largely become volatile and are constantly removed from the reactor (Hofmann et al. 2013), but the relative humidity cannot be controlled, thus it decreases considerably at elevated temperatures. The lack of control over the relative humidity necessitates the reduction of the wood moisture content to nearly zero percent in a pre-drying step at lower temperatures. Therein, the shrinkage stresses due to wood anisotropy reach a maximum and the wood is thermally modified in an oven-dry state. In contrast, the moisture content of wood at high temperature and high pressure unsaturated steam conditions in a closed reactor system is dependent on the relative water vapor pressure (Lenth and Kamke 2001; Kubojima et al. 2003; Ishikawa et al. 2004). By applying a sufficiently high steam pressure, the oven dry state of the wood can thus be avoided even at high temperatures.

Various studies (Seborg et al. 1953; Stamm 1956; Burmester 1973; Burmester 1975; Giebeler 1983; Borrega and Kärenlampi 2008a; Ding et al. 2011b; Rautkari and Hill 2014) have found that the thermal wood degradation occurs faster at elevated pressure in closed reactor systems than at atmospheric pressure in open systems. While wood degradation in open systems is mostly determined by the peak temperature and duration (Zaman et al. 2000; Alén et al. 2002; Welzbacher et al. 2007), additional parameters must be considered in the case of closed systems. However, in a closed system, the parameters of temperature, water vapor pressure and relative humidity are interdependent, which complicates the separation of their individual impact on wood degradation. When heating water within the treatment vessel to generate pressure, the resulting water vapor pressure is a function of the vessel temperature and the total amount of water present, thus their effects cannot be differentiated. In this study, a closed reactor system equipped with an external, heated water reservoir was used, as described by Willems (2009). In such a reactor system, the water temperature can be controlled independently from the temperature in the treatment vessel, thus enabling the free adjustment of the vessel temperature and the water vapor pressure throughout the process up to nearly saturated water vapor conditions.

It was the aim of this study to evaluate process parameters that determine the thermal degradation of wood in a closed reactor system. Therefore, European beech was thermally modified at different peak temperatures, for different peak durations and at different water vapor pressures, while aiming at constant relative humidity throughout the process. Wood mass loss was used as a marker for the degree of degradation taking place and related to the process conditions applied. Additionally, changes of wood moisture content as a result of the process conditions applied were considered and soluble degradation products in the final product were analyzed to draw conclusions about the mechanisms that cause thermal degradation in closed reactor systems.

2.2 Material and methods

2.2.1 Material

Thermal modification processes were performed using European beech (*Fagus slyvatica* L.) wood material originating from one log. Slats with dimensions of 35×65×800 mm³ and an average density of 0.63 g cm^{-3} were prepared from conventionally kiln-dried wood with an average moisture content of 11 %, while avoiding juvenile wood, major defects and large knots, as well as red heartwood. Each process was performed using eight slats.

2.2.2 Thermal modification process

The thermal modification processes were performed in a laboratory-scale treatment reactor, based on the technology described by Willems (2009). It consisted of a stainless steel vessel with a volume of 65 l that was connected to a heated water reservoir and to a gas washer via separate, controllable exhaust valves.

The treatment schedule consisted of four steps: (1) a 50 min. holding step at 50 °C under pre-vacuum (approx. 14 kPa), (2) a temperature increase at 12 °C/h, (3) a holding step at the peak temperature, duration and pressure given in **Table 2** and (4) a temperature decrease at -20 °C/h to 65 °C. The wood temperature was not recorded during the process. Preliminary measurements within the reactor system showed, however, that the wood temperature reaches the set point temperature of the treatment vessel, as long as pre-dried wood (< 15 % moisture content) is used. The pressure in the autoclave was adjusted by increasing the water temperature in the water reservoir and the release of excess pressure through the gas washer using the exhaust valves. Throughout the process, the pressure was controlled in a way that the relative humidity remained constant. The relative humidity, RH [%], was defined as described in equation 2.

$$RH = \frac{WVP}{P_{sat(T)}} \times 100\%,$$
(Eq. 2)

where WVP is the water vapor pressure, and $P_{sat}(T)$ is the temperature dependent saturated WVP. However, since the WVP could not be determined directly, a pre-vacuum (approx. 14 kPa) was applied and the total pressure in the vessel was used for the calculation of RH. Since this is only valid under the assumption of pure steam conditions, a precise control of RH was not possible. Below 0.1 MPa at the end of the process, the pressure could not be controlled by release of excess pressure due to a lack of overpressure in the vessel in relation to the gas washer. An exemplary process schedule is shown in **Figure 4**.

Table 2: Set point parameters applied during the thermal modification processes. Mean value and standard deviation of measured data are given in parentheses.

No.	Peak vessel-temperature [°C]	Maximum pressure [MPa]	Peak duration [h]	Set point RH [%]
1	150 (149.7 ±1.02)	0.14 (0.14 ±0.001)	3	30
2	150 (150.3 ±0.95)	0.37 (0.37 ±0.002)	3	80
3	150 (150.4 ±0.93)	0.47 (0.46 ±0.002)	3	100
4	160 (159.8 ±1.08)	0.18 (0.18 ±0.001)	3	30
5	160 (160.2 ±0.90)	0.48 (0.48 ±0.003)	3	80
6	160 (160.1 ±0.87)	0.61 (0.60 ±0.004)	3	100
7	160 (160.1 ±0.89)	0.48 (0.48 ±0.003)	1	80
8	160 (160.2 ±0.89)	0.48 (0.48 ±0.003)	6	80
9	170 (169.4 ±1.02)	0.23 (0.23 ±0.001)	3	30
10	170 (170.0 ±0.77)	0.62 (0.62 ±0.005)	3	80
11	170 (170.2 ±0.94)	0.78 (0.77 ±0.005)	3	100
12	170 (169.3 ±0.94)	0.23 (0.23 ±0.001)	1	30
13	170 (169.5 ±0.97)	0.23(0.23 ±0.001)	6	30
14	170 (170.0 ±1.20)	0.62 (0.62 ±0.004)	1	80
15	170 (170.0 ±0.83)	0.62 (0.62 ±0.004)	6	80
16	180 (179.2 ±0.96)	0.30 (0.29 ±0.001)	3	30
17	180 (179.3 ±0.85)	0.49 (0.49 ±0.003)	3	50
18	180 (179.8 ±0.78)	0.79 (0.78 ±0.006)	3	80
19	180 (179.4 ±1.07)	0.49 (0.49 ±0.003)	1	50
20	180 (179.7 ±0.95)	0.49 (0.49 ±0.004)	6	50

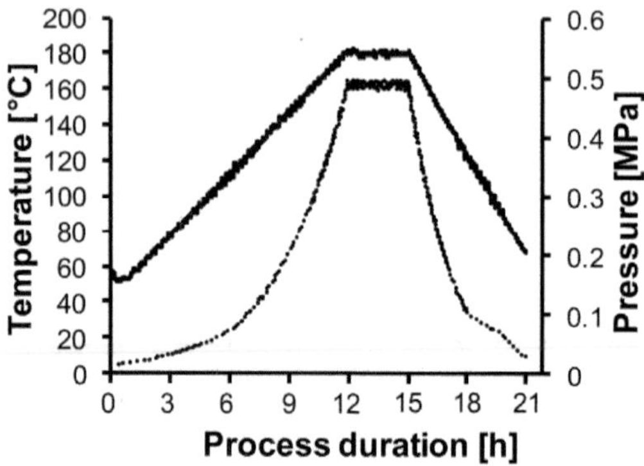

Figure 4: Exemplary process curve showing a process at a peak temperature of 180°C (solid line), a peak duration of 3 h and a maximum pressure of 0.49 MPa (dotted line) which refers to approx. 50 % set point RH.

2.2.3 Methods

In order to determine the mass loss caused by thermal modification, the weight of the slats was recorded before and after the process. Additionally, the moisture content [%], MC, was measured before and directly after the thermal modification process on small samples that were taken from each slat. MC, the dry weight of the slats and the dry mass loss of the slats during the process were calculated as described by Metsä-Kortelainen et al. (2006).

Material from each slat as well as unmodified reference material was milled and mixed in a cutting mill with a mesh size of 2 mm and used for chemical analysis. Hot water extraction was performed in duplicate for each process, using 3 g dry wood particles and 200 ml of distilled water in a Soxhlet apparatus. The amount of water soluble components, WSC, was calculated in g extractives per 100 g dry wood [g/100 g dw.] according to equation 3.

$$\text{WSC} = \frac{(w_0 - w_1) \times 100}{w_0},$$
(Eq. 3)

where w_0 is the dry weight of the wood before extraction [g], and w_1 is the dry weight of the wood after extraction [g]. Preliminary tests on selected varieties that included an extraction with ethanol-cyclohexane (1:2, by volume) after the hot water extraction resulted in an additional extractive content of less than 0.6 %. It was concluded that the majority of soluble compounds was removed by hot water, which was thus the only solvent applied during the actual test series. WSC was deducted from the dry weight of each slat to calculate the corrected dry weight, w_c [g], using equation 4.

$$w_c = w_{dry} - \left(\frac{w_{dry} \times \text{WSC}}{100}\right),$$
(Eq. 4)

where w_{dry} is the dry weight of the slat [g]. Finally the corrected mass loss during the process, ML_c [%], was calculated following equation 5.

$$ML_c = \frac{(w_2 - w_3) \times 100}{w_2},$$ (Eq. 5)

where, w_2 is the corrected dry weight of the slat before the process [g], and w_3 is the corrected dry weight of the slat after the process [g].

The soluble carbohydrate content was determined from the hot water extracts after storage in a refrigerator until analysis. The determination was carried out according to the phenol-sulfuric acid method of DuBois et al. (1956) using glucose as the standard. The soluble carbohydrate content is given in g glucose equivalents per 100 g dry wood [g/100 g dw].

For determination of the acidity, 2.5 g of dry wood particles and 30 ml of distilled water were placed on a flatbed horizontal shaker at room temperature for 24 h. The extracts were filtered and the grist was washed with distilled water to a final volume of 100 ml. Initial pH was determined using a pH meter (WTW inoLab, Type pH Level2, Weilheim, Germany). Afterwards, acidity was measured by titration using a 0.01 M sodium hydroxide solution. The amount of sodium hydroxide required to reach the neutralization point was determined from the titration curve. This amount was used as a measure for the acidity of the wood in mmol NaOH equivalents per 100 g dry wood [mmol/100 g dw]. Measurements were done in duplicate per variety.

The equilibrium moisture content (EMC) under the process conditions during the high temperature holding stage were calculated based on the assumption that MC is controlled by the water vapor chemical potential, following the explanations by Willems (2014b). After the process, small samples with dimensions of 10×10×40 mm³ (rxtxl) were prepared from each slat, resulting in 20 samples per variety. They were stored at 20 °C and 65 % RH until constant weight was reached to calculate the resulting moisture content, $EMC_{20/65}$ [%], according to equation 6.

$$EMC_{20/65} = \frac{w_4 - w_5}{w_5} \times 100\%$$ (Eq. 6)

where w_4 is the weight after conditioning at 20 °C/65 % RH and w_5 is the dry weight after drying at 103 °C for 24 h.

$EMC_{20/65}$ was used to calculate the apparent fiber saturation point, AFSP [%], according to equation 7.

$$AFSP = EMC_{20/65} \times e^{\left(R \times \frac{T}{1.39} \times \ln\left(\frac{100}{RH}\right)\right)^{0.662}},$$ (Eq. 7)

where R=0.00831 kJmol⁻¹K⁻¹ is the universal gas constant and T is the ambient temperature [K].

The AFSP was then used to calculate the EMC under process conditions, EMC_{pc} [%], following equation 8.

$$EMC_{pc} = AFSP \times e^{-\left(R \times \frac{Tmax}{1.39} \times \ln\left(\frac{100}{RH}\right)\right)^{0.662}},$$ (Eq. 8)

where T_{max} is the respective set point peak temperature [K] and RH is the respective set point relative humidity [%] during the process. Equations 6 and 7 were both derived from Willems (2014b).

2.3   Results and discussion

Due to the high-temperature and high-pressure regimes within the treatment reactor, changes in the wood can only be analyzed after the process to draw conclusions about the impact of temperature, pressure and relative humidity during the process.

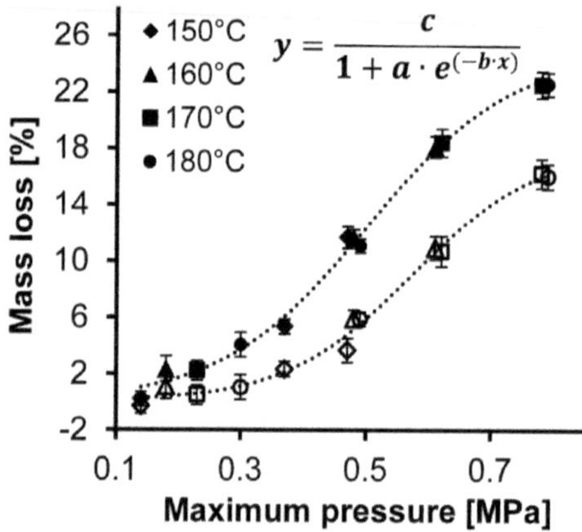

$$y = \frac{c}{1 + a \cdot e^{(-b \cdot x)}}$$

Figure 5: Mass loss [%] in dependence on the applied maximum pressure [MPa] for processes performed at a constant peak duration of 3 h. The dotted lines represent logistic functions characteristic for each data group. Open symbols: Mass loss based on dry weight (a=285.04, b=9.87, c=18.09); Closed symbols: Mass loss based on corrected dry weight (a=75.18, b=8.62, c=24.93). (N=8; ± 95% confidence interval).

Mass loss during thermal modification has been suggested to be a suitable marker for the treatment intensity of the process and is closely linked to many properties of thermally modified wood (Stamm et al. 1946; Welzbacher et al. 2007). The mass loss of wood is mainly caused by the degradation of amorphous carbohydrates and the evaporation of native extractives (Zaman et al. 2000; Alén et al. 2002). As can be seen in **Figure 5**, the mass loss is to a considerable extent affected by the pressure that is applied. Remarkably, the mass loss as well as the corrected mass loss increase with the maximum pressure, independent of the peak temperature applied. A similar observation was made by Willems et al. (2015a) for Norway spruce. It is thus concluded that the thermal degradation of wood that takes place in a closed reactor system can be best described by the pressure rather than by the temperature, within the observed range of process conditions (150-180 °C, 0.14-0.79 MPa). This finding is particularly interesting in view of the fact that the WVP was controlled independently from the temperature within the vessel, as described above. The mass loss versus pressure dependence can be described well by logistic functions. Accordingly, an almost exponential increase in mass loss to a maximum pressure of approximately 0.5 MPa is followed by saturation effects above 0.5 MPa. Such saturation effects at high mass loss levels might be assigned to the limited amount of amorphous carbohydrates within the wood to be degraded.

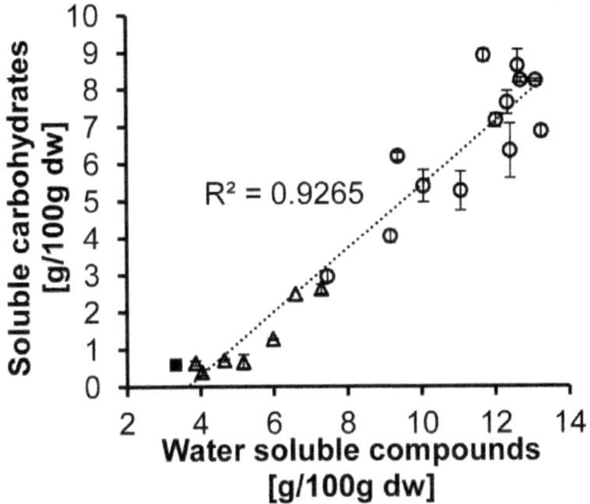

Figure 6: Soluble carbohydrates [g glucose/100g dry wood] versus percentage of water soluble compounds [g/100g dry wood.]. Square: Reference; Triangles: maximum pressure < 0.4 MPa; Circles: maximum pressure > 0.4 MPa. The dotted line represents a linear regression curve. (N=4; ± standard deviation)

However, the mass loss based on the dry wood does not consider degradation products that remain within the wood after the process. Processes at high pressure regimes lead to more than 10 % WSC. By subtracting the extractive content before and after the process to calculate the corrected mass loss, a stronger increase in mass loss with pressure can be observed. The phenol-sulfuric acid method was used as a measure for the amount of carbohydrate degradation products present in the wood after the process (see **Figure 6**), because the method is sensitive to saccharides as well as to furfural derivatives that might have formed from cell wall carbohydrates by dehydration reactions (DuBois et al. 1956). As can be seen in **Figure 6**, the amount of soluble carbohydrates increases linearly with increasing total amount of WSC in the hot water extracts, reaching up to 8.9 g glucose/100g dry wood as a maximum (170°C, 0.78 MPa). Higher amounts of WSC in wood thermally modified in saturated steam compared to wood treated in superheated steam at atmospheric pressure were also found by Karlsson et al. (2012), with carbohydrates and furfural as the main compounds. It is thus concluded that degradation products of the cell wall carbohydrates, which are gasified when performing thermal modification processes under (sub-) atmospheric pressure, accumulate within the wood during processes at elevated pressure. Therefore, WSC must be deducted from the dry mass loss of wood when studying the treatment intensity and degree of degradation that has taken place under elevated pressure regimes. This was already stated by Obataya et al. (2002) when comparing the mass loss of dry- and moist-heated spruce wood. Furthermore, high amounts of residual degradation products in the final product should also be considered during the later application, since they might have adverse effects, e.g. in terms of VOC emission, discoloration of coating systems, or mold growth. It should also be noted, that a fraction of the degradation products might have become insoluble by reacting with the remaining cell wall polymers during the thermal modification process. Typical examples of such reactions are condensation reactions that involve lignin, furfural, hydroxymethylfurfural or other reactive intermediates (Tjeerdsma et al. 1998; Garrote et al. 1999). These reactions are not accounted for when describing thermal degradation of wood by the (corrected) mass loss.

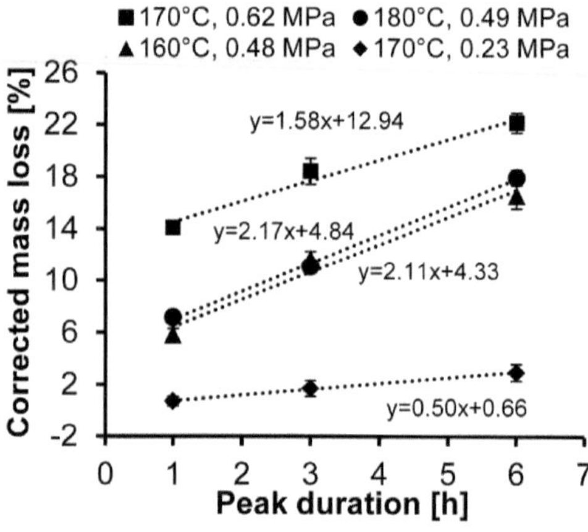

Figure 7: Mass loss [%] (corrected) in dependence on the peak duration [h] applied for different temperature and pressure regimes. Dotted lines represent linear regression curves. (N=8; ± 95% confidence interval)

The dependence of the corrected mass loss on the peak duration (**Figure 7**) can be described by linear functions. The y-intercept can be considered to be the mass loss that occurs during heating up and cooling down outside of the high temperature holding stage. It increases strongly with the maximum pressure. Due to constant heating and cooling rates, the peak temperature might have a slight additional impact by prolonging the duration of heating up and cooling down. Thermal degradation is thus not limited to the high-temperature holding stage. As suggested by Willems et al. (2015a), the slope of the mass loss versus peak duration curves can be interpreted as the mass loss rate. This mass loss rate increases from 0.5 weight-%/h at 0.23 MPa and 170 °C to 2.3 weight-%/h at 0.48 MPa and 160 °C. Processes performed at a similar pressure (0.48 and 0.49 MPa), but different peak temperatures (160 and 180 °C) result in very similar mass loss rates (2.3 and 2.2 weight-%/h). Thus, both the y-intercept and the slope point at the pressure to be the predominant factor in describing wood degradation in a closed reactor system. The lower mass loss rate of 1.6 weight-%/h at 0.62 MPa and 170 °C does not fit with this conclusion, but this might also be caused by saturation effects at mass losses exceeding 20 %, because amorphous carbohydrates are already substantially degraded. This outcome agrees with results by Seborg et al. (1953). They showed that increased pressure and/or initial moisture content lead to an elevated mass loss rate, but a stable mass loss limit was reached over time, irrespective of pressure or initial moisture content. This mass loss limit was rather dependent on the peak temperature that was applied in the observed temperature range between 205 and 300 °C.

Several studies (Seborg et al. 1953; Burmester 1973; 1975; Borrega and Kärenlampi 2008a; Ding et al. 2011b; Torniainen et al. 2011; Rautkari and Hill 2014) have consistently shown that thermal degradation is much faster for processes at elevated WVP than in superheated steam at atmospheric pressure, but different theories exist that might explain the acceleration of degradation reactions at elevated pressure. Giebeler (1983) explains the impact of pressure by improved heat-transfer in a compressed gas atmosphere which leads to an increase in the duration of the heat bath. Furthermore, there is the theoretical consideration that drying of the wood during the process and consequent evaporative cooling lowers the wood temperature within the vessel towards the dew point temperature of the water vapor (Willems et al. 2015a). Both explanations would imply that the temperature is the decisive factor for wood degradation. However, during investigations involving thermal modification of beech at atmospheric pressure in superheated steam, where evaporative cooling can be excluded due to the high-temperature pre-drying to the oven-dry state, the recorded mass loss based on dry weight for temperatures up to 180 °C did not exceed 3 % in the case of peak durations between 2 and 3 h (Kol and Sefil 2011) or 8 % for peak durations up to 36 h (Welzbacher 2010). In this study, mass losses based on the uncorrected dry weight above 10 % were even measured for peak temperatures as low as 160 °C, when using elevated pressure and peak duration of 3 h. Therefore, although improved heat transfer and reduced evaporative cooling might contribute to accelerated degradation reactions to some extent, temperature can be excluded as the sole factor for wood degradation in this study.

The compressed gas atmosphere is unlikely to directly affect the chemical reactions that occur within the wood, because drying and the formation of volatile compounds should lead to a gas flow out of the wood. However, the gas pressure in the closed reactor should determine the vaporization of liquids and organic compounds depending on several factors, such as their boiling point, the temperature as well as the composition and the absolute level of the total gas pressure. Besides the accumulation of carbohydrates and furfural shown in **Figure 6**, high concentrations of organic acids might thus build up in wood treated in closed reactor systems. The most prevalent acids in thermally modified wood have been identified as acetic and formic acid (Garrote et al. 2001; Sundqvist et al. 2006). Acetic and formic acid have an atmospheric boiling point at ca. 118 and 101 °C, respectively, and are thus rapidly vaporized and emitted from the wood during thermal modification processes in open systems (Hofmann et al. 2013; Altgen et al. 2014). In a closed system, the surrounding gas pressure should strongly affect the vaporization of organic acids and might even determine their concentration within the wood (Stamm 1956; Willems et al. 2015a). Acidic catalyzed hydrolysis and dehydration of amorphous carbohydrates have been suggested to play a major role during hydrothermal degradation of lignocellulosic materials (Fengel and

Wegener 1984; Garrote et al. 1999), and might therefore be important reactions during thermal modification in the high-pressure reactor system used in this study.

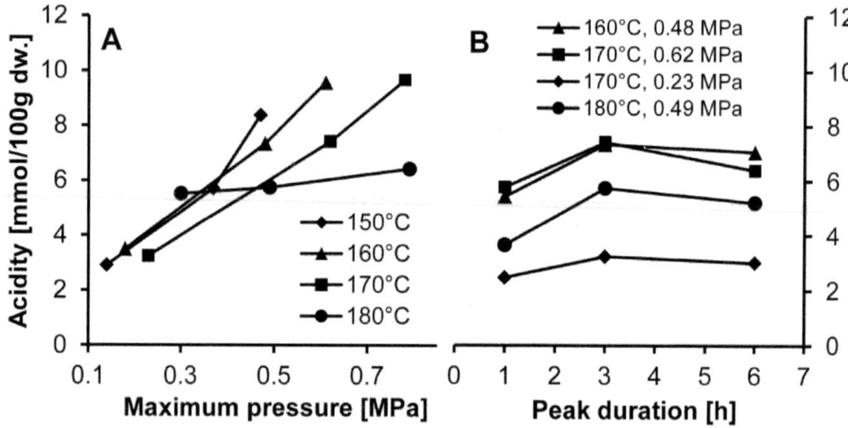

Figure 8: Acidity [mmol/100g dry wood (dw.)] of cold water extracts. A) In dependence on the maximum pressure for different peak temperatures at a constant peak duration of 3 h; B) In dependence on the peak duration for different temperature and pressure regimes.

The organic acid concentration within the wood after the thermal modification process is estimated by the acidity of cold water extracts shown in **Figure 8**. Acidity strongly increases with the maximum pressure that is applied during the process from 1.5 mmol/100 g dry wood for the unmodified reference up to 9.7 in case of beech modified at 170 °C and 0.78 MPa, while decreasing at constant pressure with increasing peak temperature. The course of acidity with increasing peak duration (**Figure 8** B) might indicate a finite acid concentration. Acidity increases between 1 and 3 h of peak duration, but remains almost constant between 3 and 6 h. The pressure-induced increase in the acid concentration might thus indeed contribute to the observed pressure-dependent increase in mass loss and mass loss rate. For a peak temperature of 180 °C, however, the impact of pressure is very low, leading to a narrow acidity range between 5.5 and 6.4 mmol/100 g dry wood for maximum pressure between 0.3 and 0.79 MPa. Accumulation of organic acids is thus unlikely to be the sole cause for the pressure-dependent mass loss progression. Treating small samples of Scots pine sapwood (*Pinus sylvestris* L.) at 150 °C and 0.47 MPa (100 % set point RH) did indeed not result in significant differences in the resulting mass loss between samples impregnated (30 min at < 130 mbar) with water and samples impregnated with 1 M aqueous acetic acid (unpublished observation by the authors).

The high-pressure reactor system used in this study aimed at controlling the WVP and thereby determined the RH and the wood moisture content in dependence on the temperature. It has previously been shown that sufficient WVP at elevated temperatures prevents complete drying of the wood, and might even enable an equilibrium moisture content in the wood during high temperatures (Lenth and Kamke 2001; Kubojima et al.

2003; Ishikawa et al. 2004). In this study the wood MC during the high temperature holding stage could not be determined. Using the final MC directly after the process to estimate the EMC during the high temperature holding stage suffers from additional effects, such as moisture consumption and production by hydrolysis and dehydration, modification of the wood and the resulting change in sorption behavior, as well as drying of the wood during the cooling phase. Instead, the EMC of the wood at room conditions (20 °C, 65 % RH) were recorded after the process and transformed to the respective set point temperature and WVP during the high temperature holding phase, using the scaling law described by Willems (2014b) to calculate the EMC_{pc}. Nevertheless, it must be noted that the release of excess pressure leads to a small deviation from the set point pressure and that residual air and volatile degradation products formed contribute to the total pressure within the vessel and thus lead to an overestimation of the WVP and the EMC_{pc}. On the other hand, frequent releases of polluted excess pressure and the generation of fresh steam in the water reservoir during the high-temperature holding phase ensure an almost pure steam environment.

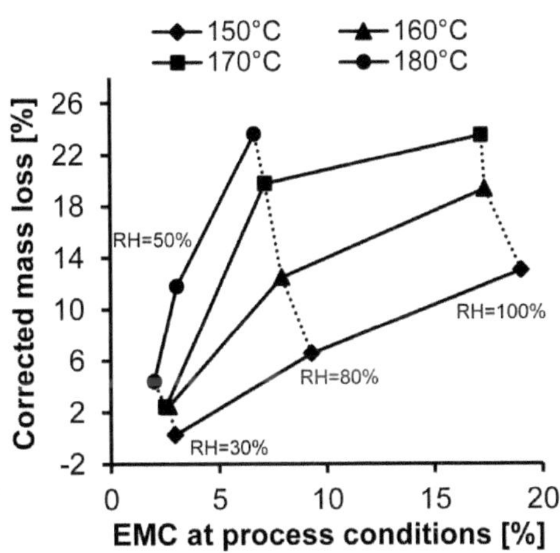

Figure 9: Corrected mass loss [%] in dependence on the calculated equilibrium moisture content at process conditions [%] for all processes performed at a peak duration of 3 h. Identical peak temperatures are connected by solid lines, while identical set point RHs are connected by dotted lines. (N=8)

Figure 9 shows the dependence of the corrected mass loss on the EMC_{pc}. It becomes evident that EMC_{pc} is mostly dependent on the RH rather than on the peak temperature. Furthermore, mass loss increases with the EMC_{pc}, with this increase being dependent on the applied peak temperature. At 150 °C, increasing the mass loss from 0.2 at 0.14 MPa (30% RH) to 13 % at 0.47 MPa (100 % RH) was associated with a strong rise in the EMC_{pc} from 2.9 to 19 %. In contrast, even though the mass loss increased from 4.4 at 0.3 MPa (30 % RH) to 23.6 % at 0.79 MPa (80 % RH), the EMC_{pc} only increased from 2 to 6.7 % when applying a peak temperature of 180 °C.

The impact of water in the wood during the process on the accelerated degradation reactions can, besides the generation of hydronium ions by water autoionization that act as catalysts for the hydrolysis of lignocellulosic materials at elevated temperatures (Garrote et al. 1999), be explained by increased thermal softening of wood. Chow and Pickles (1971) found that oven-dry samples start to soften at 180 °C with this softening increasing up to a maximum at 380 °C. An increase in wood moisture content decreased the softening temperature, and softening of wood at a constant temperature of 160 °C increased almost linearly with increasing moisture contents up to about 30 %. As hypothesized by Borrega and Kärenlampi (2008a), increased softening and mobility of wood polymers at elevated WVP might facilitate chemical reactions, lower the activation energy required and thus enable thermal degradation even at low temperatures. The impact of wood softening in the presence of water on wood degradation in closed reactor systems cannot be directly related to the WVP, but rather to a combination of temperature and RH. At constant pressure, lower temperatures might be compensated for by higher RH and thus higher wood moisture contents that reduce the activation energy required for thermal degradation. This would imply that the temperature-independent correlation between mass loss and maximum pressure shown in **Figure 5** is indirect. Since the pressure can be described as a function of RH and temperature in pure steam conditions (by transposing equation 2), it is impossible to differentiate between an effect of WVP and a combined effect of RH and temperature on wood degradation.

However, corrected mass losses above 10 % can be found in a broad variety of different EMC_{pc} values, ranging from nearly dry (3 % at 180 °C and 0.49 MPa) to severely moist (19 % at 150 °C and 0.46 MPa) conditions, which leads to the conclusion that the presence of moisture is not the only cause for accelerated wood degradation at elevated pressure. Giebeler (1983) used nitrogen to increase the pressure in a closed reactor system and found that mass loss based on the dry weight of beech increased with the applied pressure, e.g. for a temperature of 225 °C and pressure regimes between 2 and 10 bar. This is particularly interesting since the total pressure built up by nitrogen and volatile degradation products did not exceed the saturation pressure of water, hence the moisture within the wood should be rapidly vaporized in the absence of water vapor and not be a major factor for wood degradation.

Overall, none of the factors that have been discussed with regard to accelerated degradation reactions in closed reactor systems can be directly related to the observed, pressure-dependent mass loss progression. It is thus concluded that thermal degradation in the closed reactor system is, rather, caused by a combination of several factors that act simultaneously. Further studies are required, i.e. to evaluate whether the total pressure or the partial water vapor pressure is decisive for accelerated degradation reactions in closed reactor systems.

With respect to industrial implementation, pressure as the predominant process parameter in closed reactor systems to describe mass loss as a marker for thermal degradation, rather than temperature, offers potential advantages. A uniform pressure in the vessel is established automatically, thus optimal air circulation to ensure homogeneous wood degradation is not as critical in high-pressure reactor systems as it is in many temperature-dependent processes at (sub-) atmospheric pressure. High-pressure reactor systems might also enable sufficient wood degradation to change technological-relevant wood properties at low temperatures and hence at low energy consumption. However, since wood degradation was evaluated on the basis of wood mass loss and carbohydrate degradation, it remains unclear if degradation and reorganization reactions in the lignin are affected by elevated pressure as well. While there are studies indicating that degradation reactions in the lignin are activated thermally, but not catalyzed by acid (Ibbett et al. 2011), other studies state that lignin depolymerization and repolymerization even occur during treatments in saturated steam conditions as a consequence of the acidity created during the treatment (Li et al. 2007). Further studies are thus required to evaluate chemical reactions within the lignin during thermal modification of wood in high-pressure reactor systems.

2.4 Conclusions

Thermal degradation in a closed reactor system, estimated by wood mass loss, was found to be strongly determined by the maximum pressure, rather than the peak temperature applied. By increasing the pressure, the rate at which mass loss is generated increased until reaching saturation due to the limited amount of amorphous carbohydrates to be degraded. Elevated pressure also prevented excessive vaporization of degradation products (i.e. carbohydrates, furfural and organic acids) from the wood, which should be considered when using wood mass loss as a marker for thermal degradation. Different factors explaining accelerated degradation reactions have been discussed. However, the observed mass loss progression could not be assigned to one single factor, but might rather be regarded as a combination of several factors that function simultaneously. The application of steam pressure offers the possibility for a less energy consuming and more homogeneous modification process compared to processes at high temperatures and (sub-) atmospheric pressure.

Chapter 3   Publication II

Influence of process conditions on hygroscopicity and mechanical properties of European beech thermally modified in a high-pressure reactor system

Michael Altgen und Holger Militz*

Wood Biology and Wood Products, Burckhardt Institute, Georg-August University Göttingen, Büsgenweg 4, 37077 Göttingen, Germany

corresponding author:
Phone: +49-551 3933541
Fax: +49 551 399646
E-mail: hmilitz@gwdg.de

Originally published in:
Holzforschung
Walter de Gruyter Verlag
ISSN (Online) 1437-434X
ISSN (Print)   0018-3830
DOI: 10.1515/hf-2015-0235

Received: 04 November 2015 / Accepted: 03 March 2016 / Published online: 18 April 2016

Abstract: European beech (*Fagus sylvatica* L.) was thermally modified in a closed reactor system under various process conditions. Sorption cycles, dynamic vapor sorption measurements, and a three-point bending test were performed on thermally modified wood (TMW) to assess hygroscopicity and mechanical properties. As a function of mass loss (ML), the initial equilibrium moisture content (EMC) measured at 20 °C/65 % RH directly after the process was strongly influenced by the relative humidity (RH) during the process. This effect is explained by realignments of amorphous polymers in the cell wall ultra-structure in the course of thermal modification. However, the EMC of TMW gradually increased after sorption cycles consisting of conditioning over liquid water and water-soaking. This increase was most distinct for TMW modified at low RH, which is an indication for reversible ultra-structural realignments. Results of the bending test suggest that structural realignments also hindered the plastic flow of amorphous cell wall polymers, thereby reducing inelastic toughness and inelastic deflection, while other bending properties were solely affected by ML alone. Process conditions in a closed reactor systems have a profound impact on resulting wood properties and thus the partial reversibility of these property changes need to be considered during the application.

Keywords: dynamic vapor sorption (DVS), equilibrium moisture content (EMC), hygroscopicity, mass loss, relative humidity, static bending test, thermally modified wood, water vapor pressure

3.1 Introduction

The principle of exposing wood to elevated temperatures in order to alter its properties has been industrially implemented in various commercially applied thermal modification (TM) processes (Militz and Altgen 2014). Most of these TM processes operate at atmospheric pressure with superheated steam in an open reactor system. In such reactors, wood mass loss (ML) is mostly influenced by the peak temperature and its duration (Zaman et al. 2000; Welzbacher et al. 2007). Under these conditions, the volatile degradation products are emitted from the wood (Hofmann et al. 2013). By considerably exceeding 100 °C at atmospheric pressure, the relative humidity (RH) decreases severely and leads to strong drying of wood. In closed reactors, on the other hand, the elevated vapor pressure prevents excessive drying (Engelhardt 1979; Lenth and Kamke 2001) and the evaporation of degradation products (Karlsson et al. 2012; Altgen and Militz 2015). Moreover, the thermal degradation reactions are much faster (Stamm 1956) and the resulting ML can be described as a function of the maximum vapor pressure, independently of the process temperature (Altgen and Militz 2015; Willems et al. 2015a)

TM reduces the hygroscopicity and improves the dimensional stability (DS) (Stamm and Hansen 1937) as well as the biological durability (Kamdem et al. 2002). TM treated non-durable wood species become suitable for exterior applications. However, concurrent with these improvements, TM woods (TMWs) suffer from a loss in mechanical strength and ductility (Kubojima et al. 2000; Boonstra et al. 2007a), which limits their structural applications. Understanding the underlying effects of TM on the property changes is therefore essential for process optimization with regard to an optimal compromise between advantages and disadvantages. TMW properties are mainly a function of wood ML (Welzbacher et al. 2007; Chaouch et al. 2013), although the same ML and treatment intensity can be generated through a vast number of different process parameter combinations (Borrega and Kärenlampi 2008a; Candelier et al. 2013). The chemical changes in the cell wall (hydrolysis, dehydration, decarboxylation, cross-linking between cell wall polymers and degradation products) have a large share on property changes of TMW (Tjeerdsma et al. 1998).

Few studies investigated the impact of vapor pressure and RH during TM in closed reactor systems. Comparing TM at atmospheric and elevated vapor pressure, Ding et al. (2011a) found that Mongolian pine modified at elevated pressure shows a lower hygroscopicity and a higher DS, but there were no significant differences in modulus of elasticity (MOE) and modulus of rupture (MOR). Investigating TM in a closed reactor at saturated water vapor pressure, Rautkari and Hill (2014) showed that increased levels of initial moisture content (MC_{ini}) resulted in higher ML, but also in less effective improvements in DS and hygroscopicity. Borrega and Kärenlampi (2008b; 2010) found that, besides thermal degradation, mechanical properties and hygroscopicity of Norway spruce thermally modified in a closed reactor were additionally affected by drying at high temperatures. They hypothesized that wetting followed by strong drying at 150-170 °C resulted in the formation of irreversible hydrogen bonds within the wood cell wall which further reduced hygroscopicity of TM Norway spruce. However, it has recently been demonstrated that a considerable proportion of the reduction in hygroscopicity by TM is reversible (Obataya and Tomita 2002), which should be considered for application and analysis.

Altgen and Militz (2015) performed TM in a high-pressure reactor system and evaluated the influence of peak temperature, TM time, vapor pressure, and RH on ML. The present study is a continuation of this work and the intention is to test the hypothesis that besides ML the hygroscopicity and mechanical properties of TMW are also affected. The hygroscopicity of the solid material should be evaluated in repeated sorption cycles. The soluble degradation products should also be investigated as a function of hygroscopicity based on dynamic vapor sorption of milled samples with or without preceding hot water extraction. Mechanical properties should be assessed in a three-point bending test at a constant ambient humidity. Properties of TMW will be analyzed as a function of ML. More precisely, the question will be in focus whether the parameters leading to ML influence the properties beyond the summative parameter ML.

3.2 Material and methods

3.2.1 Material

European beech (*Fagus slyvatica* L.) wood originating from one log was investigated. Slats with dimensions of 35×65×800 mm^3 and an average density of 0.63 g cm^{-3} were prepared from conventionally kiln-dried wood with an average moisture content (MC) of 11 %, avoiding juvenile wood, major defects, and large knots, as well as red heartwood. Each process was performed with eight slats.

3.2.2 Thermal modification (TM) process

The TM process was performed in a laboratory-scale reactor, in which a stainless steel vessel is connected to an external water reservoir and a gas washer via steerable valves, as described by Willems (2009). The TM consists of four steps: (1) a 50 min. holding step at 50°C in pre-vacuum (< 14 kPa), (2) a temperature increase at 12 °C h^{-1}, (3) a holding step at the peak temperature, duration and pressures in **Table 3**, and (4) a temperature decrease at -20 °C h^{-1} to 65 °C. The vapor pressure was regulated by increasing the temperature in the external water reservoir and a frequent release of excess pressure to ensure a constant relative humidity (process RH), except for a pressure below 0.1 MPa at the process end. Process RH was calculated according to Willems (2009):

$$RH = 100[P_{steam}/P_{sat}(T)], \tag{Eq. 9}$$

where P_{steam} is the vapor pressure (MPa) and $P_{sat}(T)$ is the temperature dependent saturated vapor pressure (MPa). P_{steam} was estimated via the total pressure, based on the simplified assumption of pure steam conditions within the vessel.

Initial and final MC were determined on small pieces collected from each slat to calculate the dry weight of the slats directly before and after each process according to Metsä-Kortelainen et al. (2006). Additionally, material from several slats of each process as well as from unmodified reference material was milled and mixed in a cutting mill and the amount of hot water (HW) extractives was determined with 3 g dry wood particles and 200 ml deionized water in a Soxhlet apparatus. The extractive content was subtracted from the dry weight of each slat before and after each process. The mass loss (ML) of each individual slat was thus calculated based on the dry and HW extractive-free weight of the wood before and after the process. However, the wood material to be tested was neither dried nor extracted prior to the modification process or the analysis.

Table 3: Maximum temperature (T_{max}), maximum pressure (P_{max}), peak duration and process RH applied during the thermal modification processes as well as the respective amount of hot-water (HW) extractives and the final moisture content (MC) recorded after the process.

T_{max} (°C)	P_{max} (MPa)	Dura-tion (h)	Process RH (%)	HW (%)	Final MC (%)
150	0.14	3	30	3.86	3.63
150	0.37	3	80	7.29	4.89
150	0.47	3	100	12.70	6.48
160	0.18	3	30	4.65	3.46
160	0.48	3	80	10.06	4.54
160	0.61	3	100	12.41	6.28
160	0.48	1	80	7.45	4.34
160	0.48	6	80	13.26	3.41
170	0.23	3	30	5.17	2.44
170	0.62	3	80	13.12	3.84
170	0.78	3	100	11.70	5.47
170	0.23	1	30	4.04	3.03
170	0.23	6	30	5.98	1.80
170	0.62	1	80	12.34	4.61
170	0.62	6	80	12.62	3.30
180	0.30	3	30	6.60	1.93
180	0.49	3	50	9.37	1.52
180	0.79	3	80	12.03	3.27
180	0.49	1	50	9.16	1.77
180	0.49	6	50	11.06	1.68

3.2.3 Sorption test on solid samples

Tests were conducted with samples of $10 \times 10 \times 40$ mm³ ($r \times t \times l$) and ten replicates per process. The samples were first conditioned at 20 °C/65 % RH to determine the initial equilibrium moisture content (EMC_{ini}) directly after the process. Subsequently, two sorption cycles were conducted with each cycle consisting of (1) storage at 20 °C and < 2.5 kPa over deionized water and (2) vacuum drying at 20 °C and < 2.5 kPa in a desiccator and finally (3) conditioning at 20 °C/65 % RH in a climate chamber. For each step (1-3), the samples were conditioned until the weight change was less than 0.1 % (w/w) 24 h^{-1}. After the 2nd cycle, the samples were oven-dried for 24 h at 40, 60, 80, and 103 °C to determine dry weight. Subsequently, water-soaking was conducted in 200 ml deionized water at room temperature (r.t.) for each ten replicates. The samples were first vacuum-impregnated with water (20 min, 13 kPa) and water was changed daily in the course of two weeks. After water-soaking, the samples were first dried at r.t. for two weeks, followed by vacuum-drying at 20 °C and < 2.5 kPa in a desiccator, conditioning at 20 °C/6 5% RH in a climate chamber, and finally oven-drying for 24 h at 50, 70, and 103 °C to determine dry weight after water-soaking. The weight change was again determined for each step after the mass change was lower than 0.1 % (w/w) 24 h^{-1}.

EMC at 20 °C/65 % RH prior to water-soaking, EMC_1 (%), was calculated for each cycle according to Eq. 10.

$$EMC_1 = 100\left[(w_u - w_{dry1})/w_{dry1}\right] \qquad \text{(Eq. 10)}$$

where w_u is the weight of the sample after conditioning (g) and w_{dry1} is the dry weight of the sample prior to water-soaking (g). EMC at 20 °C/65 % RH after water-soaking, EMC_2 (%), was calculated according to Eq. 11.

$$EMC_2 = 100\left[(w_u - w_{dry2})/w_{dry1}\right] \qquad \text{(Eq. 11)}$$

where w_{dry2} is the dry weight after water-soaking (g).

To eliminate the impact of the additional weight of HW extractives, the corrected EMC, EMC_c (%), was calculated according to Eq. 12.

$$EMC_c = EMC(1 + E/100) \qquad \text{(Eq. 12)}$$

where E is the respective amount of HW extractives related to the dry weight of the wood before extraction (%) and EMC is either EMC_1 or EMC_2.

3.2.4 Sorption test on wood particles (via DVS)

Isotherm analysis of wood particles subjected to different RH was studied in a Dynamic Vapor Sorption (DVS) apparatus (Surface Measurement Systems, London, UK). Material from several slats per variety was milled and mixed in a cutting mill. For each variety, one measurement was performed with non-extracted and one with extracted wood. Besides unmodified reference material, TM beech (170 °C, 3 h, 0.62 MPa) was analyzed. This variety contained one of the largest amounts of HW extractives measured, which was thus chosen for the evaluation of the effect of extractives on the wood sorption. Approx. 25 mg wood particles per variety were submitted to DVS measurements at 20 °C. RH first decreased to 0 % to determine the initial dry weight of the sample, before increasing in the following sequence to 20, 40, 60, 80, and 95 % followed by a decrease to 0 % in the reverse sequence. The RH remained constant until the weight change per minute was less than 0.0008 % min^{-1} over a 10 min period and the equilibrium weight at each RH step served for EMC calculation. Adsorption and desorption were measured in three successive cycles. The EMC of non-extracted samples was corrected as described in Eq. 12.

3.2.5 Mechanical properties

A three-point bending test based on DIN 52186: 1978 was conducted on a universal testing machine (Zwick Roell Z010, Zwick, Ulm, Germany) on samples with dimensions of $10\times10\times180$ mm^3 (r×t×l) that were conditioned at 20 °C/65 % RH prior to testing. A minimum of 20 samples per variety were tested. The span length was set to 150 mm and the load was applied in the t-direction with a testing speed adjusted to cause the failure of the sample within 90 ± 30 s. Failure was defined as a load decrease of 10 % or more of the maximum load.

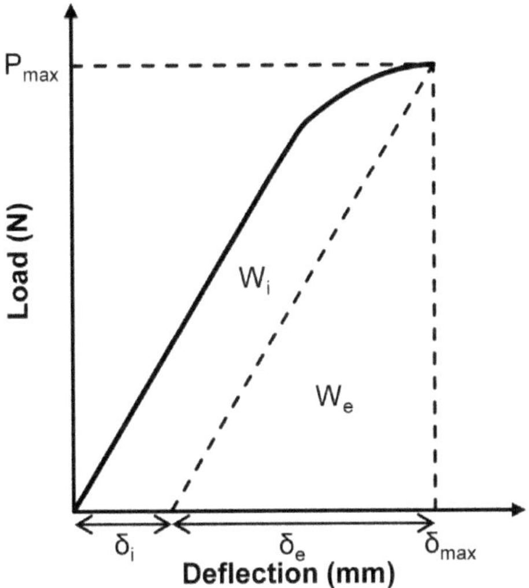

Figure 10: Schematic load-deflection curve of a three-point bending test showing maximum load (P_{max}), deflection at maximum load (δ_{max}), elastic deflection (δ_e), inelastic deflection (δ_i), elastic proportion of work in bending (W_e) and inelastic proportion of work in bending (W_i).

Modulus of elasticity, MOE [N mm^{-2}], and bending strength, MOR [N mm^{-2}], were determined as described in DIN 52186: 1978. The deflection at maximum load, δ_{max} (mm), was divided into elastic deflection δ_e (mm) and inelastic deflection δ_i (mm), as shown in **Figure 10** and Eq. 13.

$$\delta_e = (P_{max}\delta)/P \qquad \text{(Eq. 13)}$$

where P_{max} is the maximum load (N), P is any load (N) below the proportional limit, δ is the deflection (mm) at load P.

The inelastic deflection was computed by subtracting the elastic deflection from the deflection at maximum load. The area under the load-deflection curve up to maximum load was calculated as work in bending, W_{max} (Nmm), according to Eq. 14.

$$W_{max} = \int_0^{\delta max} Pd\delta \qquad \text{(Eq. 14)}$$

The work in bending was then divided into the elastic and inelastic proportion, W_e (Nmm) and W_i (Nmm), as seen in **Figure 10**. The elastic proportion was calculated according to Eq. 15.

$$W_e = (\delta_e/2)P_{max} \qquad \text{(Eq. 15)}$$

The inelastic proportion was computed by subtracting the elastic proportion from the total proportion of work in bending. All results are presented as a ratio with the mean value of the unmodified reference set to 1.

3.3 Results and discussion

3.3.1 Hygroscopicity

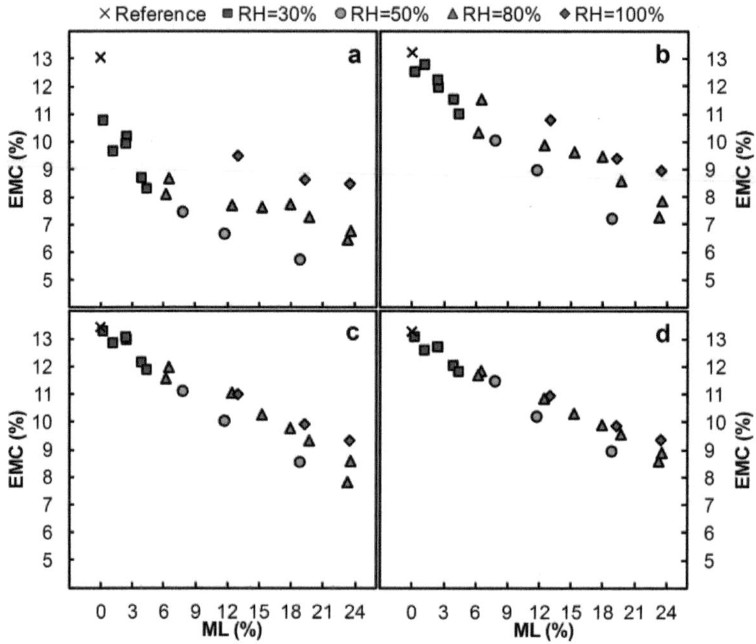

Figure 11: EMC at 20°C/65% RH (%, N=10) in dependence of the ML during the process (%, N=8). a) Initial EMC after conditioning directly after the process; b) EMC after the first sorption cycle; c) EMC after the second sorption cycle; d) EMC after water-soaking.

The $EMC_{20°C/65\%\ RH}$ was considerably affected by the process conditions. By conditioning directly after TM (**Figure 11** a), a considerable decrease in the EMC_{ini} with increasing ML was evident, leading to a minimum of 5.7 % (180 °C, 50 % RH, 6 h). However, as a function of ML, three distinct groups can be observed within the data set, which differ in the applied RH during the process: (1) $TMW_{30-50\%\ RH}$, (2) $TMW_{80\%\ RH}$, or (3) $TMW_{100\%\ RH}$. At a given ML, the EMC_{ini} of $TMW_{50\%\ RH}$ was more than 2.8 % units lower than that of $TMW_{100\%\ RH}$. Furthermore, considerable reductions in EMC_{ini} were already evident at low ML regimes. In particular, $TMW_{150°C,3h,\ 30\%RH}$ featured an EMC that was 2.3 % units lower than the reference EMC, despite an almost negligible ML of 0.2 %.

Considerable changes in the $EMC_{20°C/65\%\ RH}$ were recorded during sorption cycles (**Figure 11** b and c). While the EMC of the reference remained almost constant, it increased for TM samples in dependence on the process RH. For samples treated at low process RH, the increase in EMC during the sorption cycles was more pronounced than for those treated at high RH, thus differences in EMC between the three groups gradually diminished.

Furthermore, the EMC of samples with an almost negligible ML reached the EMC of the reference samples after the 2nd sorption cycle (**Figure 11** c). After water-soaking, the EMC vs. ML relationship was almost uniform, with only a slight residual effect of the RH (**Figure 11** d). The difference between the EMC_{ini} and EMC_{final} after water-soaking was 0.2 % units for the reference, less than 1.5% units for $TMW_{100\% RH}$, and between 2.2 and 4 % units for $TMW_{80\% RH}$ or at lower RH conditions.

It is evident that the reduction in hygroscopicity during TM processes must be separated into a reversible and an irreversible part, coinciding with observations of Obataya and Tomita (2002). Irreversible changes in hygroscopicity were present even after two sorption cycles and two weeks of water-soaking and the results were highly correlated with the ML and were not influenced much by the process conditions. Irreversible reduction in hygroscopicity can thus be assigned to the loss of hydroxyl groups caused by the degradation of hemicelluloses, although this effect might be additionally enhanced by condensation and cross-linking reactions in the lignin moiety or between lignin and degradation products of hemicelluloses (Tjeerdsma et al. 1998; Obataya and Tomita 2002). In contrast, the reversible hygroscopicity reduction was not correlated with ML, but was rather affected by the process RH. When performing TM at different process RHs, differences can be expected concerning (a) the amount of HW extractives in the final product and (b) wood drying during the process. Both factors might offer potential explanations for the observed pattern in EMC_{ini} and the reversibility of TMW hygroscopicity.

The amount of HW extractives varied considerably depending on the process conditions (**Table 3**). Large amounts of HW extractives for TM performed at high pressure regimes can be explained by limitation of excessive vaporization of degradation products leading to high residual amounts of soluble carbohydrates within the TMW (Karlsson et al. 2012; Altgen and Militz 2015). Such products might affect hygroscopicity either by their polar sites for water or by causing cell wall bulking by decreasing the volume of nanopores, which reduces sorption and swelling (Wangaard and Granados 1967; Xie et al. 2012). Higher amounts of HW extractives along with an elevated EMC_{ini} might emphasize the importance of additional polar sites in TMW on hygroscopicity. However, this interpretation cannot be linked to the hygroscopicity increment during sorption cycles and water-soaking. The latter might be explained by a cell wall bulking effect of degradation products, which is diminished after their removal. According to Biziks et al. (2015), the loss in DS of TMW during repeated water soaking cycles along with a loss of water soluble compounds should be preferably interpreted as an effect of cell wall bulking. However, ML was not observed between subsequent vacuum-drying steps that would indicate an evaporation of extractives during sorption cycles, while reversible changes in hygroscopicity were evident. Furthermore, a cell wall bulking effect is contradictory to a higher initial EMC of TMW obtained at high process RH that contained larger amounts of HW extractives.

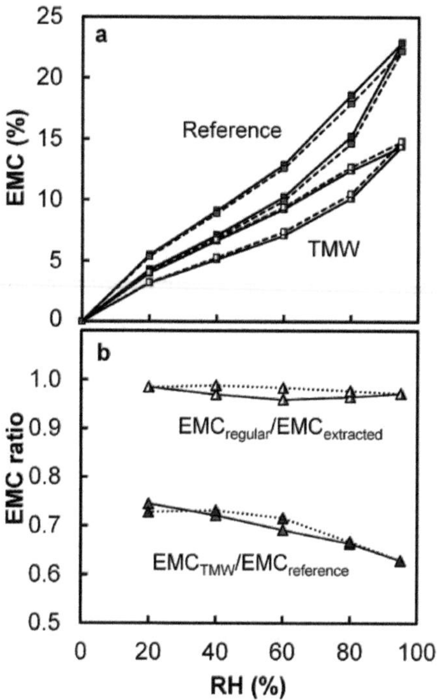

Figure 12: Results of the DVS measurements. a) Sorption isotherms of unmodified beech (filled squares) and TMW beech (semi-filled symbols) for regular particles (solid line) and extracted particles (dashed line). b) EMC ratio of regular versus extracted wood of modified beech (semi-filled triangles) and EMC ratio of modified versus unmodified beech for regular wood (filled triangles). Solid and dotted lines represent EMC ratios based on adsorption and desorption, respectively.

The impact of residual degradation products was further evaluated in the DVS experiments of TMWs with extractive contents of 3.3 % and 13.1 %, respectively. The first adsorption isotherm of extracted samples showed a reduced hygroscopicity and an increased hysteresis compared to the subsequent adsorption isotherms, presumably as a result of oven-drying of the wet particles after extraction. Reproducible isotherms were measured beginning with the 2nd DVS cycle, which was analysed. Based on the 2nd DVS cycle, HW extraction did not have a considerable impact on the sorption isotherms of the reference or TMW compared to the data of non-extracted wood, while a considerable reduction in hygroscopicity by TM was evident (**Figure 12** a). The latter was further evidenced by the EMC ratio of modified vs. unmodified wood (**Figure 12** b) that was not only reduced considerably at an RH of 20 %, but also further decreased with increasing RH in the DVS. This progressive decrease in EMC ratio is similar to the behavior of wood treated with a cross-linking agent (Himmel and Mai 2015), although far less pronounced. In contrast, the EMC ratio of non-extracted vs. extracted TMW only shows a very slight and almost constant reduction over the measured RH range with a minimum of 0.96. Accordingly, the impact of remaining degradation products on hygroscopicity of TMW is minor, and thus other factors must be taken into account in case of the reversible hygroscopicity reduction.

The effects of wood drying in this context can be limited even at high temperatures by applying a sufficiently high vapor pressure. A dependence of wood sorption on RH has been verified for T > 150 °C (Engelhardt 1979; Lenth and Kamke 2001). The MC_{final} of TMW measured directly after removing the samples from the treatment reactor is in line with the results of these studies, as the MC_{final} decreases with increasing peak temperature and particularly with decreasing process RH (**Table 3**). However, a MC decrement from 11 % even at a set point 100 % RH indicates that MC changes during TM will also be influenced by slight deviations from the set point parameters, generation or consumption of water in chemical reactions, drying during the cooling stage, when the wood temperature exceeds the process parameters, or reduction of polar sites (Altgen and Militz 2015). Nevertheless, lowering the process RH at constant peak temperature facilitates wood drying during TM. Severe drying at low process RH is a potential cause for an additional reduction in EMC_{ini} after the process. After thermal modification of Norway spruce in a closed reactor system, Borrega and Kärenlampi (2010) observed that the EMC at constant ML was ca. 2% units lower if wetting a re-drying of wood occurred compared to processes without drying. The quoted authors suggested that this effect is caused by an irreversible hydrogen bonding and structural realignments of the cell wall polymers.

Structural realignments that affect swelling and sorption in case of pulp fibers during drying are called hornification (Minor 1994). Hornification-like effects also occur probably during kiln-drying or TM of solid wood (Hillis 1984; Borrega and Kärenlampi 2010). Lowering the distance between cell wall polymers during drying contributes to the formation of additional hydrogen bonds between them (Griffin 1977; Salmén and Fahlén 2006). Such ultra-structural realignments are further facilitated by enhanced polymer mobility at temperatures higher than the glass transition temperature (Tg), where amorphous cell wall polymers change from a stiff and glassy to a soft and rubbery state (Chow and Pickles 1971; Hillis 1984). When the T falls below the Tg during the cooling phase of TM, the structural realignments become fixed (Hillis 1984). Consequently, amorphous wood polymers are restrained from swelling and the water-available polar sites are diminished as evidenced by the decrease in EMC_{ini}. Increasing the RH during TM not only prevents excessive drying and complete closing of micro-voids, but also results in a higher MC of wood during the cooling phase. Water acts as a softening agent and reduces the Tg of amorphous cell wall polymers (Chow and Pickles 1971; Salmén and Olsson 1998), which remain in a soft and rubbery state, and as such have more time for structural realignments during the cooling phase under high process RH conditions.

A reversible hygroscopicity reduction of TMW was already described under wood conditioning at a RH of 95 % or higher (Obataya and Tomita 2002; Hill et al. 2012b). By raising the RH above 60-75 % RH at r.t. strong swelling and softening of hemicelluloses occur (Vrentas and Vrentas 1991; Engelund et al. 2013). This enables stress-relaxation as well as re-opening of micro-voids within the cell wall along with a cleavage of hydrogen bonds that result in a re-accessibility of polar sites for water. Soaking the samples in water has a very similar if not even stronger effect, as it creates rapid, full water saturation and maximum swelling of the wood cell wall.

3.3.2 Mechanical properties

The bending test was performed after conditioning at 20 °C/65 % RH directly after the process, thus the samples were in the same state as presented in **Figure 11** a. However, the obtained data variation was higher when compared with the hygroscopicity results, presumably because mechanical properties are more sensitive to raw material properties, such as grain angle, density or microscopic defects.

Remarkably, the MOR ratio did not decrease severely but varied between 0.9 and 1.1 for a ML below 10 % (**Figure 13** a). The MOR ratio decreased only above 10 % ML as a function of ML and reached a minimum of 0.72. Loss in MOR at low ML levels might be compensated by a decrease in EMC (Borrega and Kärenlampi 2008b; Arnold 2010) or by an increase in compressive strength in longitudinal direction (Boonstra et al. 2007a) compared to the reference. The ML dependence of MOR at higher treatment intensities coincides with the distinct effect of hemicelluloses degradation. This process not only leads to a distribution of stresses over less cell wall material, but more importantly, it removes hemicelluloses as the interfacial coupling agent that enable stress transfer between the lignin-rich cell wall matrix and the reinforcing cellulose microfibrils (Winandy and Lebow 2001; Hughes et al. 2015). In contrast to MOR, no considerable reduction in MOE was found at all. MOE ratio increased to a maximum of 1.14 in the ML range between 5 and 20 % and decreased to its original value for higher ML (**Figure 13** a). Stiffness of the wood, although slightly increased, is therefore less sensitive to TM than strength, which is in line with previous studies (Kubojima et al. 2000; Boonstra et al. 2007a; Borrega and Kärenlampi 2008b).

No clear impact of the process RH was found for MOE or MOR, despite a lower EMC_{ini} for TMW at low process RH. Although less moisture sensitive, MOR and MOE of TMW was increasing with decreasing MC below the FSP, similar to the effect of MC of the reference (Arnold 2010). However, in case of the reference with low MCs, mechanical properties may not increase as expected and even decrease after reaching a maximum value. Unmodified wood becomes more rigid at low MCs and stress concentrations cannot be transferred to adjacent proportions of the wood (Skaar 1988). The same might account for TMW and this is a potential explanation for the absence of an effect of the initial EMC on MOR at higher ML levels. Furthermore, micro-defects might be formed during TM at low RH due to severe drying, which counteract a beneficial effect of a lower initial EMC.

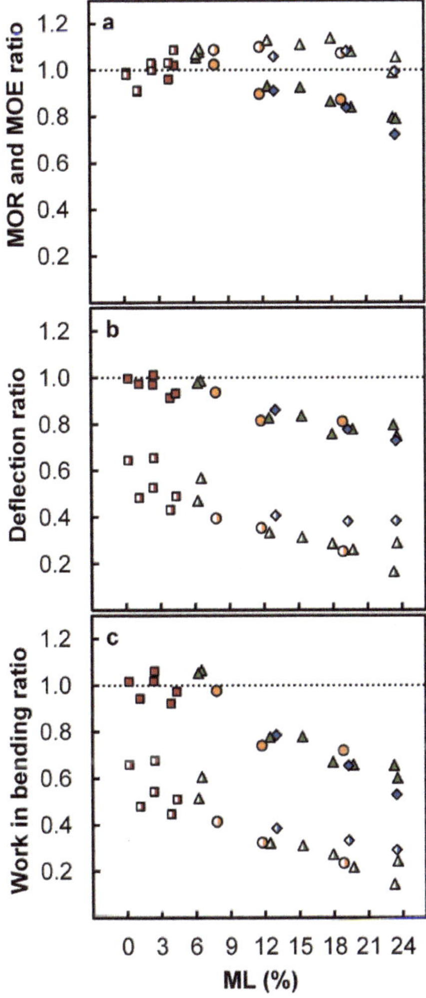

Figure 13: Results of the three-point bending test (N>20) in dependence of ML (%, N=8). a) MOR (closed symbols) and MOE (open symbols); b) Elastic (filled symbols) and inelastic (semi-filled symbols) deflection; c) Elastic (filled symbols) and inelastic (semi-filled symbols) proportions of work till maximum load in bending. Symbols are the same as in **Figure 11**. The dotted line highlights the reference value (y=1).

Deflection was separated into its elastic and inelastic proportion (**Figure 13** b). Elastic deflection ratio decreased monotonically with increasing ML, independently from the process RH during TM, to a minimum of 0.73. In line with Borrega and Kärenlampi (2008b), inelastic deflection ratio is more sensitive to TM, resulting in a minimum of 0.16. Furthermore, this property was clearly influenced by further effects. For example, inelastic deflection ratio was already considerably reduced for processes with a process RH of 30 % and low ML. In particular, a decrease to a ratio of 0.64 was evident for a process with the parameters 150 °C, 3 h, and 30 % RH, which resulted in a negligible ML of 0.2 %. Furthermore, processes with 100 % RH tended to higher inelastic deflection ratios than processes with an equal ML but at a lower process RH. Probably, the above discussed ultra-structural realignments also reduce drastically the ability of cell wall polymers for plastic flow. This resulted in high stress concentrations and failure at low deflection.

Results of the elastic and inelastic proportion of work in bending (**Figure 13** c) are a consequence of the behavior of MOR and the corresponding deflection. Similar to MOR, the elastic proportion of work in bending remained constant at a ratio around 1 for processes with a ML below 10 %, but decreased monotonically for ML higher than 10 % to a minimum of 0.53, with no clear impact of the process RH. The reduction in plastic flow and consequent inelastic deflection had a strong impact on the inelastic proportion of work in bending. A ratio below 0.68 was already evident at low ML levels for processes modified at a process RH of 30 % and the ratio further decreased with increasing ML to a minimum of 0.14. Again, processes with a process RH of 100 % tended to higher ratios compared to processes with the same ML but a lower process RH. It is concluded that structural realignments caused by low process RH reduce the inelastic toughness of the wood and cause a more brittle behavior of TMW compared to wood treated at high process RH. It is further hypothesized that a reversal of such structural realignments by exposure to high ambient RH or water soaking would increase the inelastic toughness, comparable to the results of hygroscopicity.

3.4 Conclusions

Hygroscopicity and bending properties of beech wood thermally modified in a high-pressure reactor system are considerably affected by the process conditions, i.e. the relative humidity applied during the process (process RH). As a function of ML, hygroscopicity as well as inelastic toughness decreased for wood modified at low process RH compared to thermal modification at high process RH. Ultra-structural cell wall changes and realignments of amorphous wood polymers have been proposed as a cause, while effects caused by residual degradation products within the wood were considered minor. However, the reduction in hygroscopicity was partly reversible by conditioning at high RH or by water soaking after the process, thereby diminishing the impact of process RH. The reversal of property changes has to be considered when choosing TMW for specific applications based on its properties directly after the process, as TMW properties may change under the impact of climatic conditions and time.

Chapter 4   Publication III

Wood moisture content during the thermal modification process affects the improvement in hygroscopicity of Scots pine sapwood

Michael Altgen[1]*, Tamás Hofmann[2] und Holger Militz[1]

[1]Wood Biology and Wood Products, Burckhardt Institute, Georg-August University Göttingen, Büsgenweg 4, 37077 Göttingen, Germany

[2]University of West Hungary, Faculty of Forestry, Institute of Chemistry, Bajcsy Zsilinsky u. 4, 9400 Sopron, Hungary

*corresponding author:
Phone: +49-551 3933664
Fax: +49 551 399646
E-mail: maltgen@gwdg.de

Originally published in:
Wood Science and Technology
Springer-Verlag Berlin Heidelberg
ISSN (Online) 1432-5225
ISSN (Print)   0043-7719

Received: 04 February 2016 / Accepted: 09 June 2016 / Published online: 22 June 2016

Abstract: Elevated wood moisture contents during the thermal modification process have been shown to adversely affect the improvement in dimensional stability and hygroscopicity. This study tested the hypothesis that the effect of elevated wood moisture content is based on the impact of water on chemical reactions that determine the cell wall matrix stiffness. Samples of Scots pine sapwood (*Pinus sylvestris* L.) were thermally modified in saturated water vapor at different peak temperatures and durations starting either in oven-dry or in water-saturated state. For a given mass loss caused by the modification process, the improvement in maximum swelling and equilibrium moisture content was stronger for oven-dry samples. After removal of water-soluble degradation products, which caused a cell wall bulking effect, the maximum swelling even increased after modification in water-saturated state. Based on Dynamic Vapor Sorption measurements it was evidenced that the modification in oven dry state increased the cell wall matrix stiffness which improved dimensional stability and hygroscopicity. Enhanced bond formation in the polymeric network, i.e. via condensation and cross-linking reactions during the treatment of oven-dry wood, are suggested as a cause for this increase in matrix stiffness. In contrast, the modification in water-saturated state enhanced the flexibility of the cell wall matrix, which increased the cell wall swelling and limited the improvement of hygroscopicity to the reduction in OH groups by removal of hemicelluloses. This enhanced matrix flexibility was potentially caused by predominant hydrolytic cleavage of bonds in case of water-saturated samples, evident from the chemical analysis of soluble degradation products, which increased the free volume between adjacent matrix polymers.

Keywords: dynamic vapor sorption, maximum swelling, saturated water vapor, thermal degradation, wood

4.1   Introduction

Dimensional stabilization and reduction of hygroscopicity of native wood species are some of the main benefits of thermal modification (TM) and improve the performance of wood that is subjected to exterior conditions or interior conditions with varying atmospheric relative humidity (RH). Wood responds to variations in the RH with changes in moisture content (MC). During the exposure to a constant RH, the wood will eventually reach a stable equilibrium moisture content (EMC). This hygroscopic behavior of wood is caused by hydroxyl (OH) groups as sorption sites for water. Water-accessible OH groups are mainly found in the hemicelluloses of the wood cell walls. In contrast, cellulosic OH groups are only accessible by water on the surface of the microfibrils and lignin has generally less OH groups compared to cell wall polysaccharides (Runkel 1954; Runkel and Lüthgens 1956).

When plotting the EMC against the RH at a constant temperature, the resulting curve is a sorption isotherm with a hysteresis between the adsorption and desorption isotherm. Sorption isotherms of wood and other lignocellulosic materials typically have a sigmoidal shape with a convex course at low RH and a concave course at high RH. Various models and theories have been developed over the past decades to explain this shape of sorption isotherms (Skaar 1988). In more recent studies, matrix polymer relaxation was considered to explain the sorption behavior of wood (Vrentas and Vrentas 1991; Hill et al. 2012a; 2012b; Popescu and Hill 2013; Engelund et al. 2013). Following their explanations, the adsorption of water by wood causes a swelling pressure which results in the expansion of the cell wall nanopores and thereby creates additional sorption sites within the wood matrix. This expansion is hindered by the limited free volume in the cell wall matrix and requires the rearrangement of the matrix polymers. When removing water during desorption, the nanopores, opened previously by adsorbed water, are closed, which requires the relaxation of the surrounding matrix polymers. The presence of water reduces the glass transition temperature of amorphous wood polymers. Below the glass transition temperature at low RH, the reduced polymer mobility hinders the response of the matrix to the adsorption or desorption of water and the opening and closure of the nanopores is delayed. Therefore, adsorption and desorption occur in material that is in different states. Exceeding the glass transition temperature at elevated RH enhances the mobility of the cell wall matrix polymers. This enhanced mobility results in a decrease in the hysteresis and an increase in the additional sorption sites formed upon the expansion of the cell wall nanopores.

When considering the wood-water relations in native wood, chemical changes to the cell wall matrix polymers caused by TM must inevitably affect the hygroscopicity and the dimensional stability of the wood. TM leads to a substantial degradation of amorphous cell wall polysaccharides, i.e. hemicelluloses, with the most prominent chemical reactions being: (a) hydrolysis of the molecule chain into smaller sugar units, (b) dehydration of sugar units to aldehydes, i.e. furfural and hydroxyl methyl furfural (Tjeerdsma et al. 1998; Kotilainen 2000), and (c) decarboxylation reactions that result in the formation of CO_2 and H_2O (Demirbaş 2000; Willems et al. 2013). Although being more stable against thermal degradation than amorphous cell wall polysaccharides, lignin is also affected by the cleavage of ether linkages and demethoxylation on the one hand, as well as by condensation reactions between lignin and carbohydrate degradation products or by lignin autocondensation on the other hand (Kollmann and Fengel 1965; Tjeerdsma et al. 1998; Nuopponen et al. 2005).

Reduced hygroscopicity of thermally modified wood is most commonly explained by the decreased amount of accessible OH-groups as a result of the degradation of hydrophilic hemicelluloses during the TM process. However, poor correlations were found between the recorded EMC of thermally modified wood and the amount of accessible OH groups determined by the deuterium exchange method (Rautkari et al. 2013). This leads to the conclusion that further mechanisms need to be considered. An alternative mechanism, based on the consideration of matrix polymer relaxation, was introduced by Hill et al. (2012b). They argued that TM increases the cell wall modulus and that this increase in matrix stiffness further restricts the expansion of the cell wall nanopores, leading to reduced sorption. Furthermore, additional mechanisms that are not directly caused by chemical changes, such as hornification as a result of severe drying (Borrega and Kärenlampi 2010) or cell wall bulking effects caused by remaining degradation products (Biziks et al. 2015; Čermák et al. 2015), need to be considered, as they can affect the hygroscopicity and the dimensional stability of thermally modified wood.

The thermal degradation reactions are strongly determined by the process conditions, such as the peak temperature and peak duration, but also the presence of water. Water is involved in many of the chemical reactions described above. While water is consumed as a reagent in hydrolysis reactions, it is generated as a product in dehydration, condensation or esterification reactions. Water also acts as a softening agent and increases the polymer mobility, which may reduce the activation energy required for the thermal degradation of amorphous wood polymers (Borrega and Kärenlampi 2008a). The elevated wood MC is thus considered as one of the main reasons for faster wood degradation during TM in closed reactor systems at high water vapor pressure compared with TM in open systems, in which wood is treated in oven-dry state (Borrega and Kärenlampi 2008a). However, elevated wood MC during the TM also results in a lower dimensional stabilization and a less effective reduction in hygroscopicity compared with the TM of wood in dry conditions (Stamm and Hansen 1937; Seborg et al. 1953). In a recent study, Rautkari and Hill (2014) found that increasing the initial MC of Scots pine sapwood, in the range between 0% and 26%, increased the mass loss after TM at saturated water vapor pressure, but also decreased the effectiveness of the dimensional stabilization and the reduction in EMC. To date, there is no conclusive explanation for this effect on the dimensional stability and hygroscopicity.

This study tests the hypothesis that the wood MC during the TM in saturated water vapor affects the change in cell wall matrix stiffness. It is expected that a reduced wood MC favors an increase in the cell wall matrix stiffness which further enhances the improvement in hygroscopicity and dimensional stability.

4.2   Material and methods

4.2.1   Material and treatment process

Scots pine sapwood with dimensions of 25×25×10 mm³ (r×t×l) and an average density of 0.50 g cm⁻³ was used for the experiments. Before the treatment, all samples were first oven-dried in the following sequence: 40, 60, 90 and finally 103 °C for 24 h, respectively, to determine the initial dry mass. Subsequently, they were either kept in a dry state in a desiccator or impregnated with demineralized water before the treatment. Fifteen replicates were treated in a closed, stainless steel vessel with a volume of 3 l. The vessel was filled with 300 ml of demineralized water and the samples were placed on a stainless steel sample holder above the water surface. Vacuum (4 kPa) was applied to remove residual air and the vessel was placed in an oven. The oven temperature was stepwise increased to 60, 90 and 120 °C for 1 hour, respectively, before applying the peak temperature and the peak duration given in **Table 4**. After this high-temperature holding stage, the oven was opened and the vessel cooled down. During the treatment, the water temperature was monitored, ensuring that the set point temperature was reached.

Table 4: Set point process parameters applied during the thermal modification process

No.	Peak temperature (°C)	Theor. WVP[*] (MPa)	Peak duration (h)
1	135	0.31	5
2	145	0.42	2
3			5
4			8
5	155	0.54	5

[*]theoretical water vapor pressure based on the assumption of saturated water vapor conditions

4.2.2 Chemical analyses

Five out of fifteen replicates were milled and mixed in a cutting mill with a mesh size of 2 mm and in a rotary mill with a mesh size of 0.75 mm, directly after the process. The particles were stored in sealed plastic bags at room conditions until analysis.

For the determination of volatile compounds, samples were extracted by soaking 2.5 g of wood particles in 30 ml distilled water for 24 h at room temperature; the extract was filtered and the particles were washed with distilled water to a final volume of 100 ml. Volatile compounds (formic acid, acetic acid, furfural) were measured by high-performance liquid chromatographic separation (Shimadzu LC-20 equipment, Shimadzu Corporation, Kyoto, Japan) using a Bio-Rad Aminex HPX-87H column (Bio-Rad Laboratories Inc., Hercules, USA) at 60 °C, with 0.6 ml min^{-1} flow of 0.005 M H_2SO_4 mobile phase, and UV detection at 210-240 nm. Concentrations were given in mg 100 g^{-1} dry wood. The measurements were done in duplicate and both values were in the range of a standard deviation between 6 and 11 %.

For the determination of reducing sugar and total phenol contents, 0.2 g wood particles were extracted with 20 ml 80 % aqueous methanol solution (VWR-International, Budapest, Hungary) for 20 minutes by sonication using an Elma Transsonic T570 ultrasonic bath (Elma Schmidbauer GmbH, Singen, Germany).

Reducing sugars in the methanolic extracts were determined following the 3,5-dinitrosalicylic acid (DNS) assay (Miller 1959) with glucose as standard. The amount of reducing sugars was given in g glucose equivalents per 100 g dry wood. 3,5-dinitrosalicylic acid and glucose were purchased from Sigma-Aldrich (Budapest, Hungary).

Total phenol content was measured using the Folin-Ciocâlteu assay (Singleton and Rossi 1965) applying quercetin (Sigma-Aldrich, Budapest, Hungary) as the standard. Folin-Ciocâlteu reagent was obtained from Merck (Darmstadt, Germany). Results were expressed in g equivalents of quercetin 100 g^{-1} dry wood. Both reducing sugar content and total phenol content were measured by the use of a Hitachi U-1500 type spectrophotometer (Hitachi Ltd., Tokyo, Japan). Reducing sugars and total phenol contents were measured in triplicate.

Dry mass determination for expressing results of the chemical analyses was carried out by drying 3 g wood samples to constant weight at 105 °C in an Ohaus MB23 moisture analyzer balance (Ohaus Europe GmbH, Nänikon, Switzerland).

4.2.3 Mass loss and moisture content

For ten replicates the initial, MC_i (%), the final moisture content, MC_f (%), and the dry wood mass loss, ML_{dry} (%), was determined according to equations 16-18.

$$MC_i = 100[(w_2 - w_1)/w_1], \qquad \text{(Eq. 16)}$$

$$MC_f = 100[(w_3 - w_4)/w_4], \qquad \text{(Eq. 17)}$$

$$ML_{dry} = 100[(w_1 - w_4)/w_1], \qquad \text{(Eq. 18)}$$

where w_1 is the dry weight before the process (g) and w_2 is the weight directly before the process (g), w_3 is the weight directly after the process (g), and w_4 is the dry weight after the process (g). The samples were then impregnated with demineralized water and soluble compounds were leached with 200 ml demineralized water per 5 samples and daily water changes in the course of 14 days. After leaching, the weight after oven-drying, w_5 (g), was determined once again to determine the amount of water soluble compounds, WSC (%) according to equation 19.

$$WSC = 100[(w_4 - w_5)/w_4] \qquad \text{(Eq. 19)}$$

WSC of unmodified reference samples, WSC_{Ref}, was subtracted from the dry weight of each sample before the process to calculate the corrected initial dry weight, w_c (g), according to equation 20.

$$w_c = w_1 - [w_1 \times (WSC_{Ref}/100)] \qquad \text{(Eq. 20)}$$

The corrected mass loss, ML_c (%), was then calculated based on the dry and WSC-free weight of the samples before and after the process according to equation 21.

$$ML_c = 100[(w_c - w_5)/w_c] \qquad \text{(Eq. 21)}$$

4.2.4 Maximum swelling and equilibrium moisture content

The same ten replicates were then used for the determination of maximum swelling, S_{max} (%). The cross-sectional area was measured in oven-dry state and in water-saturated state (vacuum impregnation with deionized water for 30 min at 13 kPa, followed by storage in water for 24 h). S_{max} was determined according to equation 22.

$$S_{max} = 100[(SA_{wet} - SA_{dry})/SA_{dry}], \qquad \text{(Eq. 22)}$$

where SA_{wet} is the surface area in wet state (mm^2), SA_{dry} is the surface area in oven-dry state (mm^2). Determination of S_{max} was done before and after water leaching.

After the determination of S_{max}, the samples were first dried at room conditions for two weeks, oven-dried in the temperature-sequence described above, and then conditioned at 20 °C and 65 % RH to determine the EMC (%) according to equation 23.

$$EMC = 100[(w_u - w_s)/w_s],$$
(Eq. 23)

where w_u is the conditioned weight (g).

4.2.5 Dynamic vapor sorption

Isotherm analysis of wood particles subjected to different RH was studied in a Dynamic Vapor Sorption (DVS) apparatus (DVS advantage, Surface Measurement Systems, London, UK). After determination of the maximum swelling and EMC, oven-dry samples modified at 155 °C for 5 h and water-saturated samples modified at 145 °C for 8 h were milled and mixed in a cutting mill with a mesh size of 2 mm. These two variants were selected since they featured different ML_c but resulted in a similar EMC during the measurement of solid samples. Approximately 20 mg of wood particles were placed on the sample holder of the DVS, which was connected to a microbalance by a hanging wire and was located in a thermostatically controlled cabinet. The temperature was maintained constant at 20 °C. The preset RH first decreased to 0 % to determine the initial dry weight of the sample, before increasing the RH in the following sequence: 5, 15, 25, 35, 45, 55, 65, 75, 85 and 95 % followed by a decrease to 0 % in the reverse sequence. The RH was remained constant until the weight change per minute (dm/dt) was less than 0.002 weight% min^{-1} over a 10 minute period and the equilibrium weight at each RH step was used to calculate the EMC. For each RH step, the EMC of the modified samples were related to the respective EMC of the reference sample to calculate the EMC ratio.

4.3 Results and discussion

4.3.1 Change in MC and mass loss during the process

The TM process changed the MC of the samples. Initially oven-dry samples (MC_i 0 % ±0%; mean value ±standard deviation) took up moisture, leading to MC_f between 10 and 13 % (**Table 5**), which coincides with Rautkari and Hill (2014) and is in line with wood sorption in high temperature and high water vapor pressure environments (Engelhardt 1979; Lenth and Kamke 2001; Kubojima et al. 2003; Ishikawa et al. 2004). However, the fact that MC_f increased with the peak duration of the treatments at 145 °C may indicate that an equilibrium state was not reached within the time frame of the process. In contrast, MC_f of water-saturated samples (MC_i 140 % ±10.6 %) decreased with increasing peak temperature and duration. Free water in water-saturated samples might have been partly vaporized and contributed to the water vapor pressure in addition to the excess water added to the vessel. Furthermore, the decrease in MC_f with increasing peak duration might indicate that water was also consumed in hydrolysis reactions. Nevertheless, MC of water-saturated samples did never fall below fiber saturation during the process, thus considerable differences in MC during the modification process were evident between initially oven-dry and water-saturated samples.

Table 5: Mean values of final moisture content (MC_f), mass loss based on dry weight (ML_{dry}), amount of water soluble compounds (WSC), and corrected mass loss (ML_c). Standard deviations are given in parenthesis.

Process[a]	MC_f (%)	ML_{dry} (%)	WSC (%)	ML_c (%)
Reference	-	-	1.47 (0.21)	0 (0.21)
Oven-dry				
135/5	10.98 (1.1)	1.09 (0.11)	3.00 (0.12)	2.61 (0.16)
145/2	10.02 (2.5)	0.84 (0.08)	2.28 (0.19)	1.64 (0.22)
145/5	11.00 (1.6)	1.89 (0.13)	3.93 (0.35)	4.30 (0.34)
145/8	12.99 (3.5)	3.06 (0.22)	5.85 (0.54)	7.28 (0.61)
155/5	10.32 (3.1)	3.02 (0.14)	5.22 (0.63)	6.63 (0.68)
Water-saturated				
135/5	117.57 (10.6)	1.29 (0.42)	8.39 (0.74)	8.20 (0.87)
145/2	86.02 (6.7)	1.00 (0.26)	5.70 (0.59)	5.24 (0.46)
145/5	79.18 (7.9)	1.59 (0.46)	11.50 (0.75)	11.58 (0.71)
145/8	74.85 (5.8)	2.10 (0.38)	14.55 (1.06)	15.06 (1.12)
155/5	75.33 (11)	3.27 (0.39)	15.98 (1.06)	17.43 (1.05)

[a] Peak temperature (°C)/peak duration (h)

Remarkably, the presence of water did not influence ML_{dry} considerably, although an increase in ML_{dry} with the peak temperature and duration was recorded (**Table 5**). However, the TM resulted in an increase in the amount of WSC. This increase was considerably higher for the water-saturated samples than for the oven-dry samples. Thermal degradation products strongly accumulate within the wood during TM at elevated water vapor pressure (Altgen et al. 2016), which is why the WSC was deducted from the dry mass of the samples to calculate the ML_c. The ML_c is considered to be a more accurate measure of the degree of thermal degradation reactions at elevated water vapor pressure (Altgen et al. 2016). The ML_c of water saturated samples was more than twice as high as that of the oven-dry samples. The presence of excessive amounts of water might facilitate the hydrolysis of hemicelluloses into smaller fragments, which are subsequently removed by water-leaching after the process.

The chemical analyses of soluble degradation products support this explanation (**Table 6**). The amount of reducing sugars not only increased with the peak temperature and duration but was also much higher for water-saturated samples compared with oven-dry samples. Hydronium ions generated by water autoionization catalyze the hydrolysis of hemicelluloses (Garrote et al. 1999). In a similar fashion, the total phenol content of oven-dry samples increased in dependence on the peak temperature and duration applied, but it was also found to be slightly higher in the case of water-saturated samples. The thermal cleavage of lignin and/or hemicelluloses-lignin bonds which increases the solubility of lignin in organic solvents (Nuopponen et al. 2005) might thus be further enhanced by hydrolytic cleavage in the presence of water. Elevated wood MCs should also facilitate the splitting of acetyl groups of the hemicelluloses and the formation of acetic acid (Tjeerdsma and Militz 2005).

However, the amount of acetic acid was found to be lower in water-saturated samples. Acetic and formic acid are soluble in water and become increasingly volatile at elevated temperatures. Together with the vaporization of water, the acids might have thus been released into the gaseous phase of the treatment reactor. Such evaporation might also explain the lower contents of furfural in water-saturated samples. However, it is more likely that the formation of furfural by dehydration of hydrolyzed sugars was reduced in the presence of high amounts of water as a product of such reactions. Carboxylic acids and furfural are possible reactants in esterification (Tjeerdsma and Militz 2005) and condensation reactions (Tjeerdsma et al. 1998; Garrote et al. 1999) during the TM of wood, respectively.

Table 6: Results of the chemical analysis of soluble degradation products within the wood. Standard deviations are given in parenthesis.

Process*	Reducing sugars (g/100g)	Total phenol cont. (g/100g)	Acetic acid (mg/100g)	Formic acid (mg/100g)	Furfural (mg/100g)
Reference	1.52 (0.11)	0.12 (0.01)	24.49	10.53	1.55
Oven-dry					
135/5	2.49 (0.24)	0.36 (0.01)	121.51	7.23	3.48
145/2	1.63 (0.12)	0.36 (0.01)	88.26	12.03	3.58
145/5	2.75 (0.20)	0.47 (0.01)	159.19	10.54	5.00
145/8	4.57 (0.22)	0.56 (0.02)	227.11	10.37	5.84
155/5	4.56 (0.32)	0.62 (0.02)	200.41	6.22	5.61
Water-saturated					
135/5	5.85 (0.10)	0.49 (0.01)	39.08	12.25	2.63
145/2	2.96 (0.44)	0.39 (0.01)	40.72	25.00	1.48
145/5	8.26 (0.34)	0.60 (0.02)	95.49	12.25	1.33
145/8	7.41 (0.19)	0.66 (0.01)	166.36	18.85	2.09
155/5	9.11 (0.37)	0.65 (0.01)	84.99	13.55	1.33

*Peak temperature (°C)/peak duration (h)

4.3.2 Maximum swelling and hygroscopicity tested on solid samples

The change in dimensions by capillary water uptake, assessed by S_{max}, was considerably affected by the presence of water during the process and the water-leaching after the process (**Figure 14**). By removing degradation products during water-leaching, S_{max} increased for oven-dry and water-saturated samples compared to the measurement before water-leaching. This increase in S_{max} was found to be dependent on the amount of WSC. By the leaching of WSC, the sample dimensions were reduced (**Figure 15**). This reduction is higher for dry sample dimensions than for wet sample dimensions because nanopores previously occupied by WSC within the cell wall close in dry conditions while they are partly filled with water in wet conditions. The results thus confirm a cell wall bulking effect of degradation products on the dimensional stabilization, as suggested by Biziks et al. (2015) and Čermák et al. (2015).

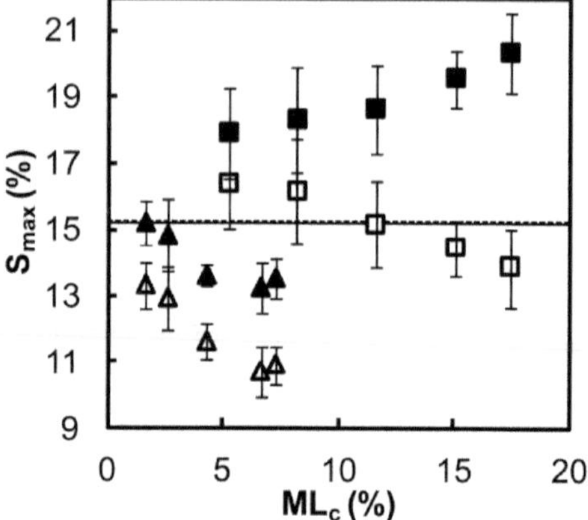

Figure 14: Maximum swelling in dependence on the corrected mass loss (ML$_c$, %). Triangles: oven-dry samples; Squares: water-saturated samples; Open symbols: before water leaching; Closed symbols: after water leaching; the dotted and the solid line represent the reference value before and after water leaching, respectively. (N=10, ±standard deviation)

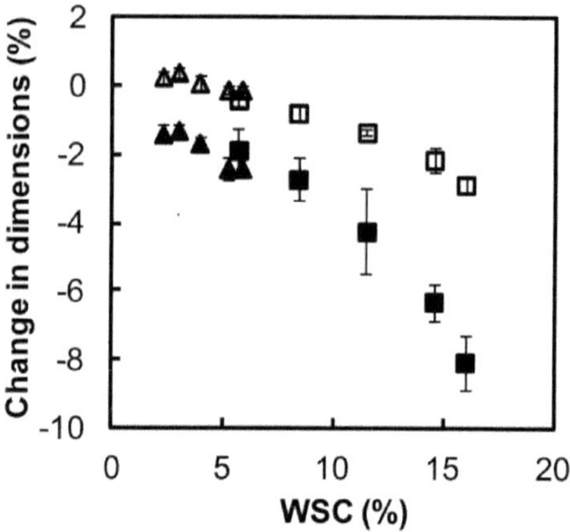

Figure 15: Change in sample dimensions by water-leaching (%) in dependence on the amount of water soluble compounds (WSC, %). Triangles: oven-dry samples; Squares: water-saturated samples; Open symbols: reduction of wet sample dimensions; Closed symbols: reduction of dry sample dimensions. (N=10, ±standard deviation)

Furthermore, it is also evident from **Figure 14** that despite resulting in a lower ML$_c$, the improvement in S$_{max}$ was higher for oven-dry samples than for water-saturated samples. This proves that the elevated wood MC adversely affects dimensional stabilization in line with Stamm and Hansen (1937), Seborg et al. (1953), or Rautkari and Hill (2014). Such an adverse effect of elevated wood MC might also explain the slightly higher S$_{max}$ for oven-dry samples treated at 145 °C for 8 h compared to those treated at 155 °C for 5 h despite a higher ML$_c$, since more moisture was taken up during a longer peak duration. Remarkably, the S_{max} of oven-dry samples still decreased after water-leaching, while S_{max} after water-leaching increased as a function of ML_c for water-saturated samples. Strong hydrolysis of hemicelluloses to mono- or oligosaccharides in water-saturated samples during TM, followed by their removal during subsequent water-leaching, greatly reduced dry dimensions but did not prevent swelling of the remaining cell wall matrix material upon water-soaking. From this it follows that there must be a mechanism other than the pure removal of hemicelluloses that reduces the swelling of the remaining cell wall matrix material evident for oven-dry samples.

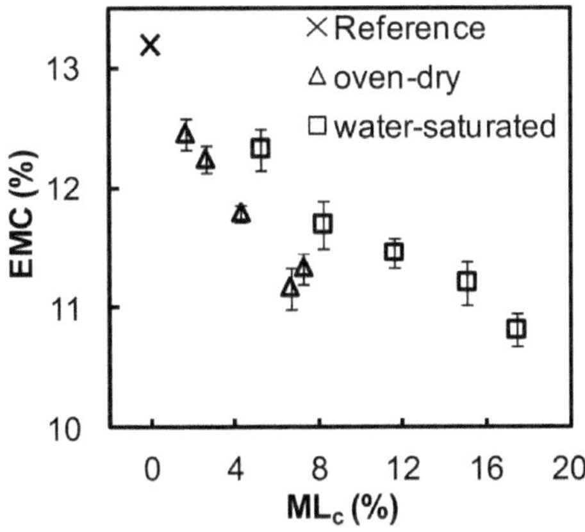

Figure 16: Equilibrium moisture content (EMC, %) measured at 20°C/65% RH after adsorption from the oven-dry state in dependence on the corrected mass loss (ML_c, %). Cross: Reference; Triangles: oven-dry samples; Squares: water-saturated samples; (N=10, ±standard deviation)

Hygroscopicity, evaluated on the basis of the EMC at 20 °C/65 % RH after water-leaching, decreased linearly as a function of ML_c (**Figure 16**). The removal of cell wall carbohydrates by hydrolysis and the subsequent water-leaching should lead to a reduction in the number of OH groups as polar sites for water. Since the removed carbohydrates are most likely those that feature the highest polymer mobility and reactivity (Borrega and Kärenlampi 2008a), accessible OH groups should be preferentially eliminated (Willems 2014a). However, the slope at which EMC decreased as a function of ML_c was higher for oven-dry samples than for water-saturated samples. Furthermore, similar to the results of S_{max}, treating oven-dry samples at 145 °C for 8 h resulted in a higher EMC despite a higher ML_c compared with a process at 155 °C for 5 h, presumably due to a higher uptake of moisture during the process. A lower EMC reduction despite a higher ML_c in the presence of water indicates that an additional mechanism other than the simple removal of hemicelluloses must apply, in accordance with the results of S_{max}.

As additional mechanism for the reduction in hygroscopicity by TM, ultra-structural realignments of amorphous polymers, denoted as hornification, have been suggested.

However, such realignments require drying during the process below fiber saturation (Borrega and Kärenlampi 2010). They can thus be excluded as an additional mechanism in this study. Instead, the results might rather be explained by an impact of the wood MC on the matrix stiffness of the wood cell wall. Accordingly, the matrix stiffness might be reduced during the TM of water-saturated samples. The presence of water facilitates the hydrolytic cleavage of bonds within the polymeric network, as evidenced by the chemical analysis. The consequent decrease in chain length and increase in free volume after the removal of hydrolyzed sugars increases the polymer mobility. The wood structure is in a more flexible and ductile state. Such a structure not only enhances swelling upon capillary water uptake but also improves the accessibility of remaining OH groups for water by the expansion of nanopores upon water adsorption.

In contrast, an increase in the cell wall matrix stiffness might have been achieved by TM of oven-dry samples. Additional bonds within the polymeric network created by condensation reactions involving, lignin, furfural and other degradation products (Tjeerdsma et al. 1998; Garrote et al. 1999) might reduce the polymer mobility. The effect of lignin to cause a stiffening of the cell wall matrix of native wood (Hosseinpourpia et al. 2016) might thus be increased. Enhanced matrix stiffness restricts the expansion of the cell wall nanopores upon water adsorption. Not only does this restriction improve the dimensional stability of wood, it also reduces the accessibility of the remaining OH groups within the cell wall and thereby decreases the hygroscopicity of wood modified in the oven-dry state. The reduced MC of oven-dry samples might have favored the formation of bonds within the polymeric network compared to the modification of water-saturated samples due to several reasons: (a) a reduced wood MC leads to a lower distance between adjacent cell wall polymers which favors cross-linking; (b) more furfural as a reactant in condensation reactions was present in oven-dry samples, as evidenced by chemical analysis; and (c) reducing the amount of water present during the TM might lead to a shift in the equilibrium of chemical reactions according to Chatelier's principle and thus facilitates condensation and esterification reactions.

A direct verification of such a shift from preferential hydrolytic splitting under wet conditions to an enhanced bond formation under dry conditions would require more advanced analytical methods, such as NMR or Raman spectroscopy. However, thermal degradation of lignin in the presence of heat and water vapor has previously been shown to involve a competition of depolymerization and repolymerization reactions (Li et al. 2007). Furthermore, the theoretical considerations are in line with studies on wood modified according to the PLATO® process. In these studies, chemical analyses after the first treatment step performed at elevated wood MC in saturated water vapor ("hydrothermolysis step") evidenced strong splitting of acetyl groups, carbohydrate chains or ether bonds within the lignin, while cross-linking within the lignin-carbohydrate-complex and esterification reactions were observed frequently in the second treatment step performed under dry conditions ("curing step") (Tjeerdsma et al. 1998; Tjeerdsma and Militz 2005). The hypothesis of elevated wood MC adversely affecting cell wall matrix stiffening and thereby limiting the improvement in hygroscopicity and dimensional stability is also consistent with the findings of Rautkari and Hill (2014) on dimensional stability and EMC of pine thermally modified in saturated water vapor.

4.3.3 Hygroscopicity tested on wood particles

To further validate the hypothesis, DVS measurements were performed with oven-dry samples modified at 155 °C for 5 h and water-impregnated samples modified at 145 °C for 8 h.

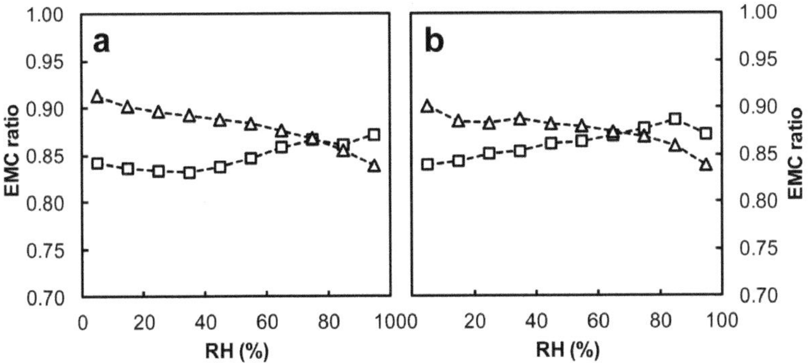

Figure 17: EMC ratio of modified wood related to the unmodified reference in dependence on the RH during the DVS measurement for adsorption (a) and desorption (b). Symbols are the same as in **Figure 16**.

The EMC ratio shows clear differences in the sorption behavior of water-saturated and oven-dry samples (**Figure 17**). At low RH levels, in particular at 5 % RH, the EMC ratio of the water-saturated sample is considerably lower than the EMC ratio of the oven-dry sample. This is in line with a higher ML_c of the water-saturated sample, which indicates a stronger removal of hemicelluloses and consequently a lower amount of OH groups as sorption sites for water. However, if the EMC of the thermally modified samples was solely based on the elimination of OH groups, the EMC ratio as a function of RH should be constant, similar to the effect of acetyl groups blocking the access to OH groups for water in acetylated wood (Thygesen et al. 2010; Himmel and Mai 2015). In contrast, a change in EMC ratio in dependence on the RH is observed for both samples. This change in EMC ratio differs between the two samples, resulting in almost identical EMC ratios in the range between 65 and 75 % RH, and in an even higher EMC ratio for the water-saturated than for the oven-dry samples above 75 % RH. The EMC ratio of the oven-dry sample decreases monotonically with increasing RH. A similar pattern, although more pronounced, is known from chemical wood modifications that are based on cross-linking, such as formalization. The enhanced cross-linking caused by formalization has been suggested to result in a stiffening of the cell wall matrix which restrains the expansion of the cell wall nanopores and thus reduces water sorption (Yasuda et al. 1994; Himmel and Mai 2015), which is in line with the effect proposed for TM of oven-dry samples.

In contrast the EMC ratio of the water-saturated sample increases with the RH within the range between 45 and 95 % (adsorption) or between 5 and 85 % (desorption). It has been suggested previously that the loss in cell wall constituents during thermal modification creates micro-voids, which enable capillary condensation of water molecules at lower levels of RH compared with unmodified wood (Hoffmeyer et al. 2003). This may explain the increase in hygroscopicity at elevated RH in case of water-saturated samples. However, capillary condensation has been found absent in unmodified wood below 99.5 % RH (Thygesen et al. 2010) and it seems unlikely that capillary condensation can explain the upwards bend in the EMC ratio of the water-saturated sample starting already at 45 % RH. Although possibly supported by capillary condensation at elevated RH, the increase in EMC ratio is more probably caused by the proposed effect of enhanced flexibility of the cell wall matrix, which further enhances the expansion of the cell wall nanopores, as suggested above.

The results of the DVS measurements thus confirm that increased wood MC during the TM process suppresses matrix stiffening but instead enhances the flexibility of the cell wall matrix. Nevertheless, it should be noted that excessive matrix stiffening, which may be achieved in high temperature, low water vapor pressure environments, limits the accessibility of sorption sites upon wood swelling or softening almost completely (Willems 2014a; 2015), but may also decrease the relaxation of stresses applied by external loads considerably (González-Peña et al. 2005). The later causes a brittle behavior with abrupt wood failures as a result of localized stresses. Therefore, a suitable compromise between stiffness and flexibility of the cell wall matrix might be required for an optimal performance of thermally modified wood. Based on the findings of this study, controlling the wood MC during the process might be the key factor to achieve such a compromise.

4.4 Conclusions

The wood MC during the TM process at saturated water vapor affects the cell wall matrix stiffness and thus determines the resulting dimensional stability and hygroscopicity. It is suggested that preferential hydrolytic cleavage of bonds within the polymeric network enhances the polymer mobility and the flexibility of the cell wall matrix when treating wood at elevated wood MC. This enhances wood swelling upon capillary water uptake and limits the improvement in hygroscopicity to the decrease in the total amount of accessible OH groups by the removal of hemicelluloses. In contrast, it is suggested that additional bond formation by enhanced condensation and cross-linking reactions under dry conditions increases the cell wall matrix stiffness. High matrix stiffness and reduced polymer mobility limit the expansion of the cell wall nanopores by adsorbed water which results in an effective reduction in hygroscopicity and swelling upon capillary water uptake.

Chapter 5   Publication IV

Wood defects during industrial-scale production of thermally-modified Norway spruce and Scots pine

Michael Altgen, Stergios Adamopoulos und Holger Militz*

Wood Biology and Wood Products, Burckhardt Institute, Georg-August University Göttingen, Büsgenweg 4, 37077 Göttingen, Germany

corresponding author:
Phone: +49-551 3933541
Fax: +49 551 399646
E-mail: hmilitz@gwdg.de

Originally published in:
Wood Material Science & Engineering
Taylor & Francis
ISSN (Online) 1748-0280
ISSN (Print)   1748-0272
DOI: 10.1080/17480272.2014.988750

Received: 8 October 2014 / Accepted: 13 November 2014 / Published online: 26 January 2015

Abstract: This research investigates wood defects, particularly the formation of surface cracks, during the production of thermally-modified wood and its exposure to cyclic moisture changes. Boards of Norway spruce and Scots pine originating from different steps within the production of ThermoWood® were collected and wood defects were investigated at macroscopic and microscopic scale. Subsequently, the wood was exposed to capillary wetting cycles to record its sensitivity towards cracking. After the modification process, typical anatomical defects of conventional kiln-drying be

,came more frequent and severe, with the magnitude being to some extent depending on the presence of defects in the raw material. At microscopic scale, damages to ray parenchyma and epithelial cells as well as longitudinal cracks within the cell walls of earlywood tracheids were evident in thermally-modified wood. Despite a lower water uptake and higher dimensional stability, thermally-modified wood was more sensitive to surface cracking during wetting cycles than unmodified wood, i.e. at the outside face of outer boards (near bark). For limiting surface cracking of thermally-modified wood during service life, the use of high-quality raw material, the exposure of the inside face of the boards (near pith) and the application of a surface coating is considered beneficial.

Keywords: Thermal modification, light microscopy, SEM, surface cracks, cell-wall features, wetting cycles

5.1   Introduction

The presence of cracks and other defects is a major drawback of wood and may reduce its service-life in exterior above ground applications. Cracking not only reduces the market value of wood; it also reduces the mechanical strength and leads to an increased water uptake, thus creating optimal moisture conditions for decay fungi attack (Chang et al. 1982; Eaton and Hale 1993). The formation of cracks is primarily related to dimensional changes caused by the hygroscopic and anisotropic behavior of wood. As a result, stresses during drying, processing and weathering of wood might occur and potentially lead to various kinds of deformations. If these stresses exceed the fracture strength of wood, cracks are developed (Lamb 1992).

During kiln-drying, surface cracks are usually formed during the early stages in case the wood surface dries too quickly. This results in tension stresses within the surface layers of wood that might exceed its tensile strength perpendicular to the grain. Wood responds to these drying stresses by creeping, which relaxes the stresses and limits cracking, although some cracks might still develop (Morén 1993). The surface cracks developed during early stages of drying or even during heating usually are very small and not visible to the naked eye, but might grow into visible deeper and longer cracks at a later stage (Hanhijärvi et al. 2003). Most problems with surface cracking during drying have been encountered with flat sawn boards, where cracks mainly occur along the rays towards the pith (Schniewind 1963; Ward and Simpson 2001).

During weathering, the wood is constantly wetted and re-dried, thus repeated swelling and shrinkage occurs. These cyclic changes in moisture content and dimensions are most pronounced at the wood surface, which is directly exposed to rain, humidity and sunlight (UV and visible light). Hence, tensile stresses occur in the outer layers, because the sub-surface layers of wood dry more slowly and thus restrain the shrinkage of the surface (Forest Products Laboratory 1957). The consequent formation of surface cracks is believed to be strongly dependent on the annual ring orientation, due to the anisotropy of the wood. Therefore, wood boards with a vertical annual ring orientation generally feature less cracks than boards with a horizontal ring orientation (Flæte et al. 2000; Sandberg and Söderström 2006). Investigations by Sandberg (1996; 1997) indicated that the pith and the surrounding juvenile wood facilitate the formation of cracks in sawn timber upon wetting and subsequent drying.

One of the most commonly applied thermal modification process in Europe is the ThermoWood® process that uses superheated steam at atmospheric pressure. The wood is dried to approximately zero percent moisture content in a high temperature drying step at the beginning and has to be reconditioned again at the end of the process, e.g. by a water spraying system (Syrjänen and Kangas 2000; Mayes and Oksanen 2003). Thermal modification of wood improves the biological durability (Kamdem et al. 2002; Welzbacher and Rapp 2002; Hakkou et al. 2006) and dimensional stability (Yildiz 2002; Popper et al. 2005), and thus aims at improving the service life of wood in exterior above-ground conditions. The application of thermally-modified wood in load-bearing constructions is, however, restricted by its strength loss (Kubojima et al. 2000; Boonstra et al. 2007a; 2007b). This deterioration in mechanical strength is primarily related to chemical changes occurring during the modification process (Tjeerdsma et al. 1998; Boonstra et al. 2007a), but defects also caused by the process are likely to adversely affect the properties of thermally-modified wood (Boonstra et al. 2006a; Boonstra et al. 2006b; Awoyemi and Jones 2011). One of the main advantages in using thermally-modified wood for exterior applications is the decreased hygroscopicity and consequently the improved dimensional stability. However, despite its lower moisture content, cracking of thermally modified wood during long-term natural weathering has been reported to be similar to that of unmodified wood (Jämsä et al. 2000). A potential reason for this contradiction could be pre-existing defects in thermally-modified wood, such as radial or tangential cracks, damaged axial tracheid walls or parenchyma and epithelial cells in rays and resin canals (Boonstra et al. 2006a; Awoyemi and Jones 2011), that act as a starting point for severe cracks caused by repeated swelling and shrinkage of wood during outdoor exposure.

To answer the hypothesis of the pre-existing defects cause, a wood anatomical investigation was performed on Scots pine and Norway spruce boards coming from different steps of the regular production of ThermoWood®. Conventional kiln-drying prior to the modification process, high-temperature drying at the beginning of the process and conditioning by water spray at the end of the modification process were postulated as the most critical steps during the production in terms of cracking. Furthermore, it was tested if lowering the heating rate during the thermal modification process can limit cracking by preventing excessively fast drying and heating of the wood. Investigations were performed on a macroscopic scale as well as by means of light and scanning electron microscopy. Subsequently, the wood was exposed to several capillary wetting cycles in order to evaluate the crack formation caused by repeated swelling and shrinkage.

5.2 Materials and methods

5.2.1 Wood materials and thermal modification

Boards of Norway spruce (*Picea abies* (L.) Karst.) and Scots pine (*Pinus sylvestris* L.), originating from six different lots, were provided by the International ThermoWood Association (Helsinki, Finland). The boards were cut in a 2- or 4-ex-log sawing pattern (**Figure 18**) with the dimensions shown in **Table 7**, and thus contained a mixture of juvenile wood, sapwood and heartwood. In case of a 4-ex-log sawing pattern, inner boards contained a higher proportion of juvenile wood and heartwood, while outer boards mainly consisted of sapwood. The boards were conventionally kiln-dried below fiber saturation point at temperatures below 90 °C. The thermal modification process was performed according to the requirements of the standard ThermoWood® class "Thermo-D", with a treatment temperature of 212 °C, as described by Mayes and Oksanen (2003). For each lot a minimum of ten boards was selected in order to account for the standard production quality. Pieces, approximately 1 m in length, were cut from the selected boards after different steps of the regular production of ThermoWood®: (A) after conventional kiln-drying, (B) after conventional thermal modification schedule and, (C) after thermal modification without conditioning step (water spray).

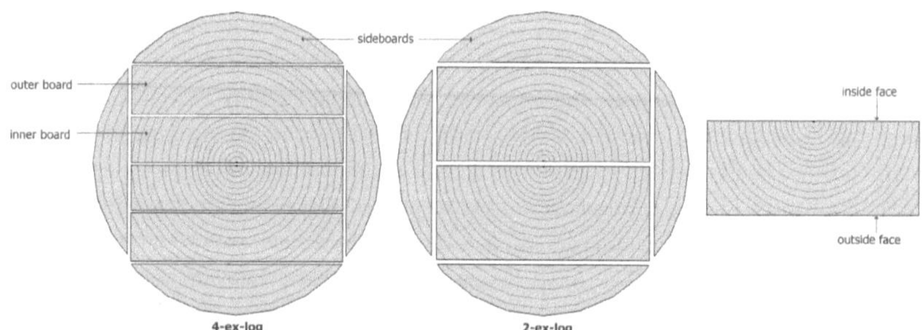

Figure 18: Sawing patterns of the boards used for the anatomical investigations. Sideboards were discarded.

Table 7: Parameters of the test material and process steps investigated for each lot in publication I.

Lot No.	Wood species	Dimensions (mm)	Sawing pattern	No. of boards	Process	Process steps[a]
1	Scots pine	50 × 100	2-ex log	30	Industrial	A, B
2	Scots pine	25 × 150	2-ex-log	30	Industrial	A, B
3	Norway spruce	38 × 125	2-ex-log	30	Industrial	A, B
4	Scots pine	35 × 150	2-ex-log	10	Industrial	A, B, C
5	Scots pine	25 × 150	4-ex-log	10	Industrial	A, B, C
6	Norway spruce	25 × 150	4-ex-log	10	Industrial	A, B, C
7	Scots pine	25 × 150	4-ex-log	10	Laboratory	A, B, D
8	Norway spruce	25 × 150	4-ex-log	10	Laboratory	A, B, D

[a] (A) after kiln-drying; (B) after conventional Thermo-D schedule; (C) after Thermo-D schedule without final conditioning; (D) after mild Thermo-D schedule; matched 1 m pieces from each board were available for all process steps

Additional material was prepared on laboratory scale with boards of Norway spruce and Scots pine that originated from five different logs per species which were cut in a 4-ex-log sawing pattern. During the study, it was differentiated between outer boards (near bark) with higher proportions of juvenile wood and heartwood and inner boards (near pith) mainly with sapwood. The material was conventionally kiln-dried at a maximum temperature between 60 and 80 °C and then it was thermally-modified according to the ThermoWood® standards using a treatment temperature of 212 °C with either an industrially applied treatment schedule or with a milder treatment schedule (step D) featuring a low heating rate. For both modification processes the treatment temperature was 212 °C and was maintained for 3 h. Details on the wood material used as well as on the thermal treatments applied can also be seen in **Table 7**.

5.2.2 Macroscopic evaluation

The matched pieces with a length of 1 m that were obtained from the full-length boards were visually evaluated and the formation of cracks was studied on the inside and outside faces (see **Figure 18**). Additionally, one cross-cut of 50 mm in length was taken from each piece for stereoscopic observation. These cross-cuts were taken from the cross section that was not directly exposed to the kiln-atmosphere during the conventional kiln-drying or the thermal modification process, thus almost 1 m from the ends of the original full-length boards. Cracks and anatomical features were investigated at magnifications up to ×10 using a S8 APO stereoscope (Leica, Wetzlar, Germany). Based on the amount and width of the cracks that were present, the inside (near pith) and outside face (near bark) of each board was rated using an evaluation scale ranged from 1 to 3 (CR=1: no defects visible, CR=2: small cracks visible only with stereoscope, CR=3: high amount of small cracks or clearly visible cracks).

5.2.3 Light microscopy

Unmodified and thermally-modified cross-cuts of Scots pine and Norway spruce with no defects visible under the stereoscope were selected for a more precise analysis by means of light microscopy. Small wood blocks with dimensions of $10 \times 10 \times 10$ mm³ were taken from the longitudinal surfaces of the selected cross-cuts using a chisel. The blocks were softened by soaking in deionized water for one night prior to microtome cutting. Transverse, radial and tangential sections (about 20 µm thick) were prepared from the wood blocks with a sliding microtome. The sections were stained with 1 % safranin solution, mounted on glass slides in glycerin, and then viewed under an Eclipse E600 light microscope equipped with a DXM 1200 digital camera (both Nikon, Düsseldorf, Germany).

5.2.4 Scanning electron microscopy

Thermally-modified cross-cuts of Scots pine and Norway spruce with no defects visible during stereoscopic evaluation were also selected for scanning electron microscopic observation. Small blocks with precise anatomical orientation and dimensions of approx. $5 \times 5 \times 5$ mm³ were prepared using a razor blade. The transverse surfaces of blocks were wetted and carefully smoothed by a sliding microtome while clear radial and tangential surfaces were carefully produced by splitting with a razor blade. The samples were mounted on aluminum stubs with carbon adhesive tape, dried overnight in an oven at 50 °C, and sputter-coated with carbon. Coated samples were observed with a scanning electron microscope (Supra 45, Leo Elektronenmikroskopie GmbH, Oberkochen, 5kV).

5.2.5 Wetting cycles

The wood pieces of lot 7 and 8 that were prepared on laboratory scale (see **Table 7**) were used after they were planed to create a smooth surface. Samples with a length of 250 mm were cut from each piece and afterwards the end-grain surfaces of the samples were sealed with a coating (Pyrotect Holz Color finish, Rütgers organics GmbH, Mannheim, Germany). The samples were exposed to three capillary wetting cycles, with each cycle consisting of a water bath for 12 h at room temperature, followed by drying in an oven at 50 °C for another 24 h. Before and after each cycle, the outer (near bark) and inner face (near pith) of each sample was evaluated and ranked for surface cracking according to ISO 4628-4: 1982.

5.3 Results and Discussion

5.3.1 Macroscopic evaluation

Based on the macroscopic evaluation of Scots pine and Norway spruce wood material coming from the regular production of ThermoWood®, it can be stated that even the kiln-dried boards (step A) that have not been subjected to the thermal modification process are not crack-free. On the inside faces (near pith), severe cracking was particularly evident when the pith was present within the wood piece. This had to be expected due to the reduction of the radius of curvature of the annual rings close to the pith, which increases the stress level during drying as a result of the shrinking anisotropy (Sandberg 1996). On the transverse sections of these pieces it could be observed that cracks started from the pith and reached the longitudinal surface (**Figure 19** A). As a result, severe longitudinal cracks could be seen on the inside face of the selected piece. Furthermore, cracks following the annual rings borders could be seen occasionally on the inside faces of the unmodified wood (**Figure 19** B). As this particular delamination type is not common in conventional drying of the species, it can be assumed that micro-cracks were already present and could had been induced during exposure of green wood by a combination of factors such as wood anisotropy, annual ring structure and juvenile wood (Bucur 2011). On the outside faces (near bark), longitudinal cracks could be observed as well (**Figure 19** C). They were mainly evident at the center of the piece, thus following the rays on the transverse section (**Figure 19** D). These later cracks represent typical defects, which can be observed after wood drying (Ward and Simpson 2001).

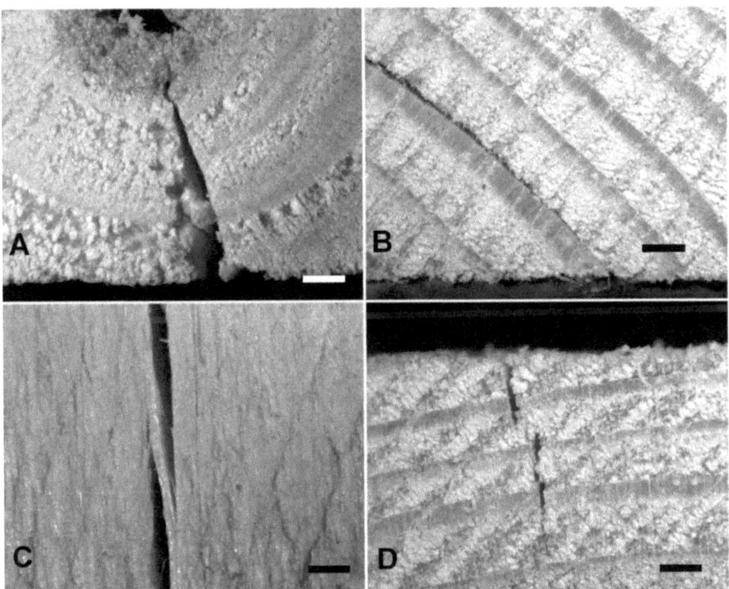

Figure 19: Typical cracks already present in kiln-dried wood as observed under the stereoscope at ×10 magnification. (A) radial crack starting from the pith on a transverse surface of Norway spruce; (B) tangential crack along the annual ring border on a transverse surface of Scots pine; (C) axial cracks on a tangential surface of Scots pine; (D) radial cracks on a transverse surface of Scots pine. Scale bars: 1 mm

The pattern of cracks observed after the ThermoWood® process did not differ from the ones described for the unmodified boards. However, based on the crack rating, it was apparent that the surface cracks on the inside (near pith) and outside faces (near bark) became more frequent and/or more severe after the ThermoWood® process (**Figure 20**). For each lot, the amount of defect-free boards (CR=1) decreased, while the amount of boards with small

(CR=2) or even severe cracks (CR=3) increased. Large variations in the crack rating existed among the boards, with no clear impact of the wood species or the board dimension. However, the presence of cracks and defects in the raw material seemed to affect the cracking of the thermally-modified wood. As an example, in lot 6 with almost 90 % of the unmodified boards being defect-free, the respective thermally-modified boards feature a comparably low amount of cracks as well, even though the amount of defect-free boards still decreased to approx. 60%. In contrast, more than 60 % of the unmodified boards in lot 3 were not defect-free, resulting in a comparably high number of thermally-modified boards with defects, i.e. at least 40 % of them with severe defects.

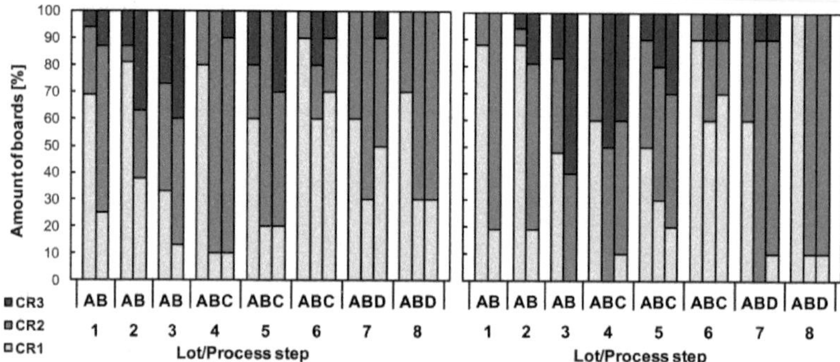

Figure 20: Percentage of boards per crack rating (%) originating from various lots and different steps of the regular production of ThermoWood®. No differentiation between inside and outside boards has been made in case of a 4-ex-log sawing pattern. Left: rating of the inside face (near pith); Right: rating of the outside face (near bark) of the pieces. (A) after kiln drying; (B) after conventional thermal modification schedule; (C) after thermal modification without final conditioning; (D) after mild thermal modification schedule with decreased heating rate; CR=1: no defects visible, CR=2: small cracks visible only with stereoscope, CR=3: high amount of small cracks or clearly visible cracks.

Severe stresses, which result in wood defects, are likely to occur during the ThermoWood® process, especially during: (a) the high temperature drying step, which decreases the moisture content to nearly zero percent before the actual thermal modification step, leading to a maximum shrinkage of the wood, or (b) the final conditioning step, where water is sprayed on the boards, creating a fast swelling of the wood at the surface (Syrjänen and Kangas 2000; Mayes and Oksanen 2003). However, macroscopic evaluation of pieces of Norway spruce and Scots pine that were not exposed to the conditioning step showed the very same defects and features as regular thermally-modified wood. Differences in the crack rating between thermally-modified wood with and without conditioning step did not follow any clear trend. In both cases, cracks were more severe and/or frequent than on the respective unmodified pieces. Therefore, the formation of cracks most likely occurred at an earlier stage of the modification process, which is either the high-temperature drying step or the subsequent treatment at 212°C. After performing the thermal modification process on laboratory scale, no considerable differences in the amount of surface cracks could be observed between the two different treatment schedules that were used. Thus, even a very low heating rate, which should prevent an excessively fast drying of the wood surface, could not prevent an increase in cracking during the thermal modification process.

Having excluded the impact of the conditioning step or the heating rate, the decrease of the moisture content to nearly zero percent and the resulting maximum shrinkage, which is unavoidable during the thermal modification with superheated steam at atmospheric pressure, is a likely cause for the increase in cracking. Besides the formation of new cracks, the increase in cracking by thermal modification might additionally be facilitated by the enlargement of closed or microscopic cracks that were already present in the kiln-dried wood. The latter suggests that special attention should be paid to the conventional kiln-drying step prior to the ThermoWood® process.

During the preparation of samples for wetting cycles by planing, the amount and the severity of surface cracks was greatly reduced. However, the higher amount and severity of surface cracks on thermally-modified wood increased the risk of residual cracks after the planing process.

5.3.2 Microscopic observations

Light microscopy revealed that defect-free material of the macroscopic evaluation might still have several defects on a microscopic level. Such micro-defects were mainly associated with the thin-walled parenchyma cells of the rays or the epithelial cells of the resin canals, especially for Scots pine. This finding is in line with the microstructural investigations of softwoods treated with the PLATO-process (Boonstra et al. 2006a). Although uniseriate rays seemed to be intact for thermally-modified Scots pine and Norway spruce, damages of parenchyma and epithelial cell walls could be observed in fusiform rays on tangential sections. On transverse sections, radial cracks along the rays were observed, especially within latewood (**Figure 21** A). Since the cell walls of the axial tracheids appeared intact, these cracks seemed to be associated with the damages of ray cells as observed on the tangential sections. Compared to tracheid walls, parenchyma and epithelial cell walls are considerably thinner and are typically not lignified in the sapwood (Tsoumis 2009). Therefore, they are less likely to withstand stresses within the wood during the thermal modification process. Furthermore, deformations of the axial tracheids (**Figure 21** C) and longitudinal cracks within the tracheid walls (**Figure 21** D) were observed, especially at the earlywood-latewood interface or within earlywood. In contrast, the thick-walled latewood tracheids appeared to be intact.

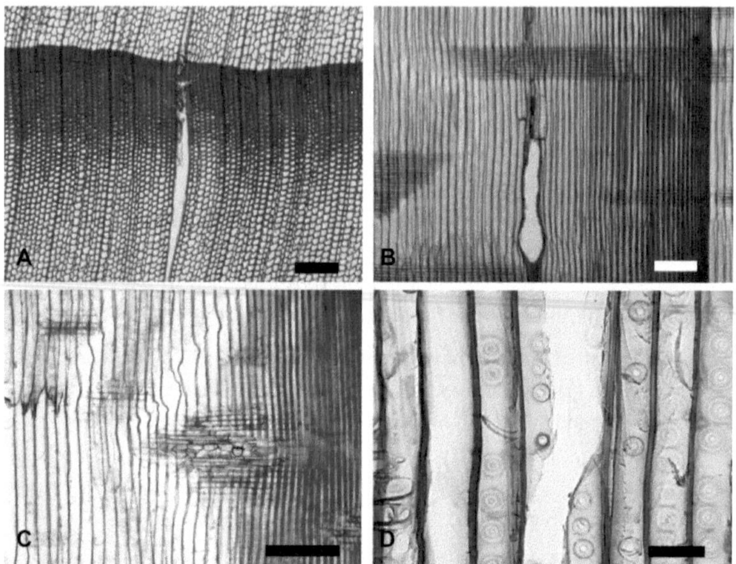

Figure 21: Typical defects observed in thermally-modified wood employing light microscopy. (A) radial crack following a ray on a transverse section of Norway spruce; (B) destructed axial resin canal on a radial section of Norway spruce; (C) disrupted earlywood tracheid walls on a radial section of Scots pine; (D) longitudinal crack in an earlywood tracheid on a radial section of Scots pine. Scale bars: 250 µm for A-C; 50 µm for D

It should be noted that with light microscopy there is a risk to artificially disturb the wood structure during sectioning and microslide preparation. This is especially valid for the thermally-modified wood, which breaks more easily due to its increased brittleness. Micro-defects in thermally-modified wood were therefore only regarded as such, if they were observed repeatedly and were not a typical feature in the unmodified references. Furthermore, scanning electron microscopy (SEM), which does require the drying of the samples under full vacuum, but not the preparation of thin slides, revealed the very same defects in thermally-modified wood as described above by using light microscopy (**Figure 22** A-C). It was thus concluded that the wood defects are present within the thermally-modified Scots pine and Norway spruce rather than being produced by the sample preparation. Furthermore, the higher resolution of SEM allowed the observation of further features in the wood microstructure. However, damaged bordered pits and a more open structure of cross-field pits along with ruptures in the pit membranes as described by Awoyemi and Jones (2011) for thermally-modified western red cedar and by Sehlstedt-Persson et al. (2006) for thermally modified Scots pine sapwood were not observed. The structure of bordered and cross field pits of thermally-modified Scots pine and Norway spruce was found to be intact (**Figure 22** D), thus confirming the findings of Boonstra et al. (2006a).

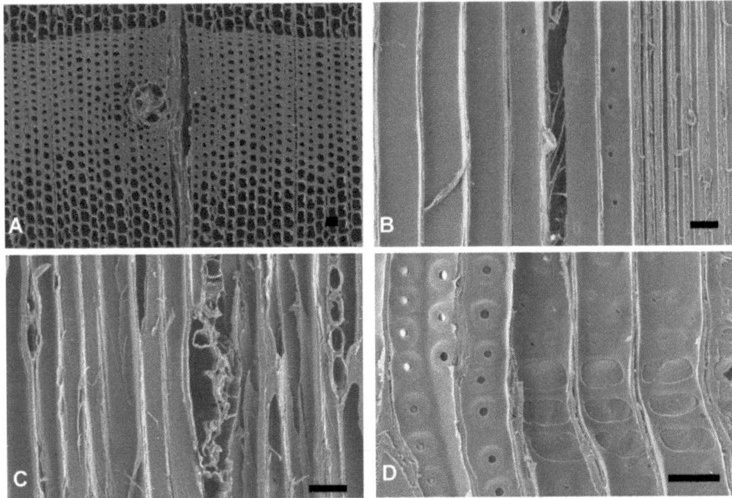

Figure 22: Anatomical features in thermally-modified wood observed with a scanning electron microscope. (A) radial crack following a ray on a transverse section of Scots pine; B) longitudinal crack in an earlywood tracheid near the earlywood-latewood intersection of a radial surface of Norway spruce; (C) destructed fusiform ray on a tangential section of Norway spruce; (D) bordered and cross-field pits in Scots pine. Scale bars: 30 μm

5.3.3 Wetting cycles

In order to imitate the formation of cracks caused by repeated swelling and shrinkage in wood used outdoors, the material was exposed to simple wetting cycles. During the wetting cycles, the mass of the thermally-modified samples did not change as much as the unmodified, kiln-dried material as shown in **Figure 23**. Thus, the water uptake of thermally-modified wood was considerably decreased by the thermal modification process, as also reported by Metsä-Kortelainen et al. (2006). With the boards including sap- and heartwood in different proportions, the change in mass of lot 7 (Scots pine) suffers from high standard deviations and a detailed analysis of the impact of pure sap- and heartwood on the water uptake is difficult. However, an increase in the capillary water absorption of Scots pine sapwood thermally-modified at mild treatment temperatures (< 200 °C), as described by Metsä-Kortelainen et al. (2006) and Johansson et al. (2006), was not observed for outer boards which should mainly contain sapwood. This deviation from previous studies might be explained by the above described absence of damages to the pit structure that have been suggested as a cause for the increased absorption of thermally-modified Scots pine sapwood (Sehlstedt-Persson et al. 2006). Furthermore, high treatment temperatures as applied in our study (212 °C) will strongly reduce the wettability and thus the resulting capillary forces. The effect of decreased capillary forces might outweigh the effect of an increased number of capillaries generated through micro-defects on the water absorption (Johansson et al. 2006). In addition to a reduced water uptake, a strong cupping of the unmodified samples was clearly visible, caused by the lower dimensional stability compared to thermally-modified wood (Yildiz 2002; Popper et al. 2005).

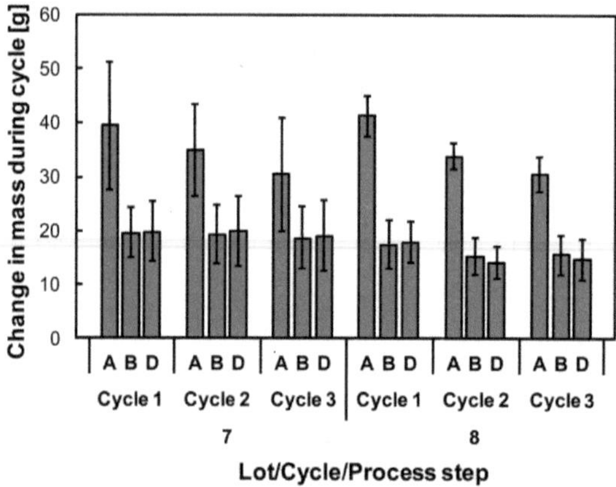

Figure 23: Average change in mass [g] during the wetting cycles as a consequence of water bath and subsequent drying for material from lots 7 and 8 originating from different productions steps. (A) after kiln-drying; (B) after conventional thermal modification schedule; (D) after mild thermal modification schedule with decreased heating rate. Each column represents a mean value of 10 samples that included different proportions of sap- and heartwood. Error bars represent standard deviation.

Despite the lower moisture uptake and better dimensional stability, the impact of the wetting cycles in terms of cracking was higher for the thermally-modified wood compared to the unmodified material. Although a milder treatment schedule seemed to decrease the cracking, in case of lot 7 compared to the application of a conventional treatment schedule, the cracking was still considerably higher than for the unmodified material.

As can be seen in **Figure 24**, the outside faces (near bark) of the boards were more sensitive to cracking than the inside faces (near pith). Although cracks following the annual rings occasionally occurred at the inside face of the boards, cracks usually developed along the rays at the center of the outside faces. Annual ring loosening on the inside face, which is a frequent problem known from outdoor applications of thermally-modified spruce, was not a typical feature for the samples exposed to the wetting cycles, potentially due to the lack of UV irradiation. During the wetting cycles outside faces of outer boards were most sensitive to cracking, despite containing a lower amount of juvenile wood, which has been found to increase crack formation (Sandberg 1996; 1997). The higher amount of sapwood in outer boards compared to inner boards might have affected the results in the case of Scots pine by means of a higher water uptake. In particular, outer boards of unmodified Scots pine suffer from severe cracking on the outside faces, which was not observed in case of unmodified Norway spruce. However, thermally-modified spruce, which does not feature distinct differences between sap- and heartwood (Metsä-Kortelainen et al. 2006), also results in a higher crack rating for outside faces of outer boards compared to inner boards. Therefore, the higher sensitivity towards cracking of outer boards was rather caused by the more horizontal annual ring orientation (tangential), in contrast to the more vertically oriented annual rings (radial) in inner boards. With the tangential shrinkage of the wood exceeding the shrinkage in radial direction (Tsoumis 2009), a more horizontal ring orientation will lead to more stresses within the wood. Furthermore, the more horizontal annual ring orientation of outer boards resulted in a higher proportion of rays that were oriented face to face. In contrast, inner boards featured a higher amount of vertically oriented annual rings and thus a higher proportion of rays that were oriented from the edges to the center of the board, and this fact might even prevented severe face-to-face cracking. This coincides with the findings of (Flæte et al. 2000) for unmodified Norway spruce and aspen after accelerated weathering.

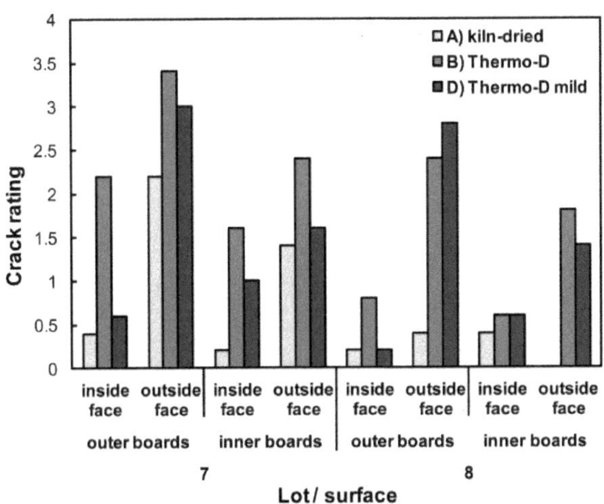

Figure 24: Crack rating according to ISO 4628-4: 1982 for lot 7 and 8 after three wetting cycles.

Thermal degradation of wood components leads to a loss in mass; therefore stresses are distributed over less cell wall material. In addition, cleavage of linkages within the wood polymeric network might hinder the distribution of stresses. Thermally-modified wood thus features lower mechanical strength (Boonstra et al. 2007a; Korkut et al. 2008a), and is more sensitive towards crack initiation and crack propagation (Majano-Majano et al. 2012). This coincides with a lower resistance against stresses caused by wetting and subsequent drying found in this study, although the absolute stress level should be lower for thermally-modified wood due to its improved dimensional stability. Furthermore, the parenchyma and epithelial cells within the rays are frequently damaged during the thermal modification process, as observed in microscopic analysis. These micro-defects might act as a starting point for more severe cracks and thus lead to a higher sensitivity towards cracking during repeated swelling and shrinkage of wood.

5.4 Conclusions

Based on the anatomical investigations of Scots pine and Norway spruce wood material originating from different steps of the regular production of ThermoWood®, the following conclusions were drawn:

- Surface cracks already present in the wood after kiln-drying became more frequent and severe after the thermal modification process.
- The parameters studied did not show an impact of the conditioning step at the end of the modification process or a milder treatment schedule, which leads to think that the high-temperature drying to nearly zero percent moisture content is the most critical factor for the crack formation during the ThermoWood® process. Nevertheless, the use of high quality raw material with only few or no defects might reduce the amount of cracks in the final product.
- Even though surface cracks can be removed by planing, damages to parenchyma and epithelial cells as well as longitudinal cracks within the earlywood tracheids could still be observed in thermally-modified wood by means of microscopy.
- Despite a lower water uptake and a higher dimensional stability, thermally-modified wood was more sensitive to the formation of cracks during wetting cycles compared to unmodified wood. A milder treatment schedule had very limited impact. Micro-defects caused by the thermal modification process might act as starting point for severe cracks during repeated swelling and shrinkage.
- During wetting cycles, the lowest number of cracks were formed on the inside face of inner boards. As long as annual ring loosening is not a critical factor, the inside face (near pith) should thus be the weathered exposed surface of thermally-modified pine and spruce in outdoor applications.

5.5 Acknowledgements

The authors thank the International ThermoWood Association (Helsinki, Finland) for their financial support, the provision of test material and fruitful discussions.

Chapter 6   Publication V

Photodegradation of thermally-modified Scots pine and Norway spruce investigated on micro-veneers

Michael Altgen* und Holger Militz

Wood Biology and Wood Products, Burckhardt Institute, Georg-August University Göttingen, Büsgenweg 4, 37077 Göttingen, Germany

corresponding author:
Phone: +49-551 3933664
Fax: +49 551 399646
E-mail: maltgen@gwdg.de

Originally published in:
European Journal of Wood and Wood Products
Springer-Verlag Berlin Heidelberg
ISSN (Online) 1436-736X
ISSN (Print)   0018-3768
DOI: 10.1007/s00107-015-0980-3

Received: 13 April 2015 / Accepted: 20 October 2015 / Published online: 26 October 2015

Abstract: The exposure to ultraviolet light results in surface photodegradation of uncoated wood during exterior application. When using thermally-modified wood in exterior applications, chemical changes that occur during the modification process might affect a subsequent photodegradation. This study investigates the impact of artificial weathering on the photodegradation of thin micro-veneers of thermally-modified Scots pine and Norway spruce by means of FT-IR spectroscopy and micro-tensile strength testing. FT-IR spectra revealed photodegradation reactions of lignin that resulted in the disappearance of the peak at 1508 cm^{-1} after 144 hours, irrespective of the thermal modification process. Loss in micro-tensile strength by photodegradation was higher for finite- than for zero-span micro tensile strength as a result of the loss of amorphous carbohydrates. Although strength loss of unweathered veneers was evident for thermally-modified wood, the rate at which strength loss occurs during artificial weathering was considerably decreased by the thermal modification process. It was concluded that the amorphous carbohydrates were less degraded during artificial weathering as they were already pre-degraded during the thermal modification process.

Keywords: ThermoWood®, artificial weathering, UV-degradation, FT-IR spectroscopy, micro-tensile strength testing

6.1   Introduction

Ultraviolet (UV) light is regarded to be the primary reason for surface degradation and discoloration of wood during weathering in exterior conditions, which is the main application for thermally-modified wood. Photodegradation of wood is a surface phenomenon, because the UV-light cannot penetrate the wood deeper than 75 µm (Hon and Chang 1984). The degradation starts with the absorption of UV-light, with lignin being the main UV-light absorbing component in the wood (Norrström 1969). The subsequent degradation process occurs through radical reactions, due to the formation of free radicals in the lignin. It has been shown that a combination of UV-light and water has a greater effect on the chemical reactions than UV light alone, because of the leaching of degradation products out of the wood (Anderson et al. 1991). Even though lignin is most strongly affected by the chain scission reactions initiated by the free radical formation, the cell wall carbohydrates are affected as well (Hon and Chang 1984). The degradation mainly occurs in the amorphous regions of the carbohydrates, since the highly organized crystalline regions of the cellulose are insensitive to the UV light (Hon and Feist 1981).

For the quantification of the photodegradation of wood, micro-tensile strength testing has previously been shown to be a suitable technique (Raczkowski 1980; Derbyshire and Miller 1981; Derbyshire et al. 1995; Derbyshire et al. 1996; Derbyshire et al. 1997; Turkulin and Sell 2002). It is based on measuring the tensile strength of thin micro-veneers after exposure to natural or artificial weathering. The low thickness of the veneers allows for a nearly full penetration of the UV-light. Strength losses can therefore be attributed to degradation reactions of the cell wall components. The method has also been successfully applied to the evaluation of strength losses during chemical modification (Xie et al. 2007; Xiao et al. 2010).

During the thermal modification process, various reactions occur that change the chemical structure and composition of the wood. These changes involve the main cell wall components, i.e. by an increase in the percentage and a further cross-linking of the lignin, or by an acid-catalyzed degradation of the amorphous carbohydrates (Tjeerdsma et al. 1998; Kotilainen et al. 2000). The chemical changes lead to an increased durability and dimensional stability of thermally-modified wood (Popper et al. 2005; Hakkou et al. 2006; Esteves et al. 2007), thus enable its use in many exterior applications. However, during long-term natural or artificial weathering, the original brown color of uncoated thermally-modified wood is not stable, because photodegradation reactions take place (Jämsä et al. 2000; Nuopponen et al. 2004; Huang et al. 2012; Yildiz et al. 2013). Nevertheless, it has been suggested that photodegradation products of thermally-modified wood do not leach out as easily as for unmodified wood. Also, changes in the polysaccharide content between weathered and non-weathered samples have been reported to be less pronounced compared to unmodified wood, because the amorphous carbohydrates are already degraded during the thermal modification process (Nuopponen et al. 2004). For short durations of artificial weathering, e.g. for 72 h (Huang et al. 2012) or even up to 835 h (Ayadi et al. 2003), a higher resistance against UV light due to the chemical changes in the lignin, such as an increase in the lignin percentage or the formation of phenolic compounds during the thermal modification process, has been proposed.

In this study, we evaluated whether chemical changes during thermal modification affect the resistance of wood against photodegradation. For this reason, micro-tensile strength testing combined with FT-IR spectroscopy was applied to thin veneers of thermally-modified Norway spruce (*Picea abies* (L.) KARST.) and Scots pine (*Pinus sylvestris* L.) after artificial weathering up to 144 hours.

6.2 Materials and methods

Kiln dried boards of Norway spruce (*Picea abies* (L.) KARST.) and Scots pine (*Pinus sylvestris* L.) were thermally-modified according to the ThermoWood® standards, as described by Mayes and Oksanen (2003). The boards were four meter in length, had a two ex-log sawing pattern and dimensions of 32×125 mm² in case of pine and 32×150 mm² in case of spruce. Each board was cut in half and two meter long pieces were either thermally-modified at 190 °C for 3 h or at 212 °C for 3 h. A total of five boards per treatment was used for the tests. The residual two meter long pieces of each board were kept as unmodified references to enable the pairwise comparison between a thermally-modified piece and the respective unmodified piece.

Small wood blocks (15×30×50 mm³; r×t×l) were cut from the edges of the boards in a way that reference and thermally-modified blocks originated from the same annual rings. Pine heartwood was avoided. Two blocks per board were prepared, resulting in a total of

10 blocks per variety. They were vacuum-impregnated with aqueous ethanol (10 % v/v) and stored for 15 days. Micro-veneers with a thickness of 100 µm were cut from the radial surface using a sliding microtome. The application of a lower veneer thickness (80 µm) in preliminary tests carried the risk of occasionally causing defects to the thermally-modified veneers during handling, i.e. in wet state directly after cutting, due to the low strength and high brittleness of the material. By slightly increasing the veneer thickness to 100 µm this problem was avoided. A deviation of 10° from the real radial plane was adjusted so that the rays did not stretch over the veneer surface. After conditioning at 20 °C and 65 % relative humidity the veneers were checked for uniformity of thickness with an electronic gauge. The veneers were exposed to 0, 8 (24 h), 24 (72 h) and 48 (144 h) weathering cycles in a QUV weathering device on the basis of EN 927-6: 2008, with one cycle consisting of 2.5 h of UV irradiation followed by 0.5 h of water spray (see **Table 8**). For micro-tensile strength testing, six conditioned micro-veneers per board, treatment and weathering duration were used for zero-span testing with a clamping pressure of 90 Psi and no free clamping range. The same amount of micro-veneers was used for the finite span testing with a free clamping length of 25 mm and a testing speed of 1 mm min^{-1}. The strength was measured in N mm^{-2} and the strength loss was calculated by setting the respective unmodified, non-weathered reference to 0 %. Additionally, selected micro-veneers of each group were measured by means of FT-IR spectroscopy in ATR mode in the wavelength range between 4000 and 750 cm^{-1} with a resolution of 4 cm^{-1} and 32 scans per spectra. Spectra were baseline corrected and normalized with the highest band set to an intensity of 100.

Table 8: Weathering cycles in the QUV weathering device on the basis of EN 927-6: 2008.

Step	Function	Temperature	Duration	Condition
1	Condensation	45°C	2 h	
2	UV	60°C	2.5 h	0.89 W/(m²nm) at 340 nm
3	Spray		0.5h	6-7 l/min, UV off

6.3 Results and discussion

6.3.1 FT-IR spectroscopy

Figure 25 shows FT-IR spectra of weathered veneers after different exposure times, exemplarily for unmodified pine and pine thermally-modified at 212°C/3h. Bands that are especially sensitive to the impact of artificial weathering in case of all varieties can be found at 1508, 1465, 1265, 1235 and 870 cm^{-1}. All of these bands can first and foremost be attributed to the lignin (see **Table 9**). The considerable decrease in absorption of these bands can therefore be explained by the degradation reactions occurring in the lignin, which is known to be preferentially degraded by UV light (Hon and Chang 1984).

Table 9: FT-IR assignments of bands affected by artificial weathering

Peak	Position in cm^{-1}	Assignment	References
a)	1508±5	aromatic skeletal vibration in lignin	[1, 2, 3]
b)	1465±5	C-H asymmetric bending in CH_3 and CH_2	[1, 2, 3, 4]
c)	1265±5	C-O stretching vibration in lignin	[1, 4]
d)	1235±5	C-O stretching vibration in lignin	[3, 4]
e)	870±5	C-H out of plane bending vibration in lignin	[4]

[1]Faix (1991), [2]Pandey and Theagarajan (1997), [3]Zhang and Kamdem (2000), [4]Harrington et al. (1964)

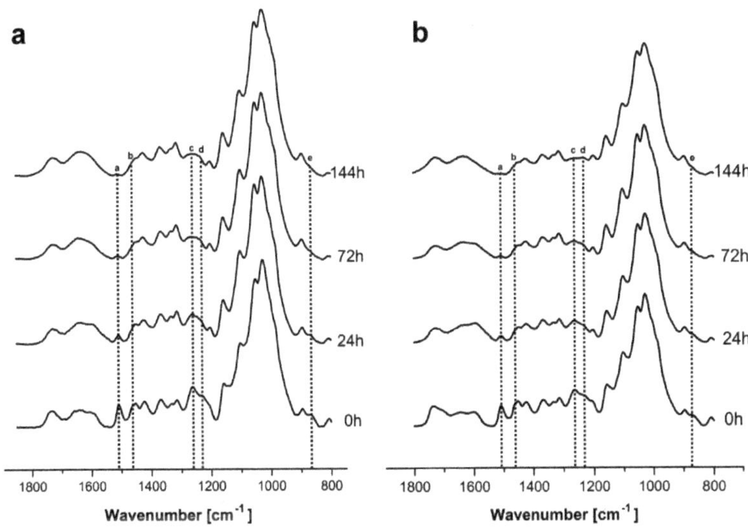

Figure 25: FT-IR spectra in the wavelength region between 1800 and 800 cm^{-1} for micro-veneers after different exposure times in the QUV weathering device. a) Unmodified Scots pine; b) Scots pine thermally-modified at 212 °C for 3 h.

For unmodified as well as for thermally-modified wood, the band at 1508 cm^{-1} disappeared after 144 h of artificial weathering, as can be seen for Scots pine in **Figure 25**. It is therefore concluded that chemical changes during the thermal modification process, e.g. cross-linking reactions or an increase in the lignin percentage, do not prevent the lignin from being degraded. Severe photodegradation reactions in the lignin of thermally-modified wood have also been reported by Yildiz et al. (2013) during artificial weathering between 400 and 1800 h, or by Huang et al. (2012) who measured a maximum lignin content of 2.5 % for thermally-modified wood after 1512 h of artificial weathering. However, contradicting with the spectroscopic measurements in our study, changes in the lignin and the formation of phenolic compounds during the thermal modification process have previously been suggested to contribute to an improved resistance against photodegradation based on color measurements on solid wood after short durations of artificial weathering such as 72 hours (Huang et al. 2012) or even 835 hours (Ayadi et al. 2003). It has also been suggested that weathering products of lignin are largely unleachable in thermally-modified wood (Nuopponen et al. 2004). Potentially, the very small dimensions of the micro-veneers used in this study facilitate the leaching of phenolic compounds and weathering products, thus diminishing their contribution to a photostability of thermally-modified wood.

6.3.2 Micro-tensile strength testing

The impact of thermal modification is apparent for micro tensile strength testing of non-weathered veneers (**Figure 26**). Micro tensile strength in both test modes, finite and zero span micro tensile strength (f-strength and z-strength), decreases considerably for thermally-modified veneers compared to the unmodified references. For conventional tensile strength testing, the adverse effect of the thermal modification is well known (Boonstra et al. 2007a; Korkut et al. 2008a). The tensile strength of wood is thought to be strongly determined by the cellulose content and its chain length (Ifju 1964). Depolymerization of cellulose occurs during the thermal modification process, but is believed to be predominant in the amorphous regions (Tjeerdsma et al. 1998; Boonstra and Tjeerdsma 2006). The resulting increase in the relative amount of crystalline cellulose might have an adverse impact on the tensile strength, because the highly ordered and rigid structure might break more easily than the more flexible amorphous cellulose (Boonstra et al. 2007a). Furthermore, Boonstra et al. (2007a) suggest that the cleavage of secondary and covalent bonds within the hemicellulosic polymer or between hemicelluloses and cellulose/lignin might interfere with the load-sharing capacities of the lignin-hemicelluloses matrix that interconnects the cellulose microfibrils. The predominant degradation of hemicelluloses during the modification process can thus be considered as a major factor for the tensile strength loss of thermally-modified wood.

Figure 26 shows that micro tensile strength loss by thermal modification is more pronounced for f-strength than for z-strength prior to artificial weathering. A treatment at 212 °C leads to an average strength loss above 40 % for f-strength, while the loss in z-strength is less than 30 % for both wood species. Similar results have been reported by Klüppel and Mai (2012) for veneers after acidic hydrolysis of amorphous carbohydrates with sulfuric acid and by Xie et al. (2007) for veneers after treatment with magnesium chloride. For conditioned veneers, they found higher losses in f-strength than in z-strength. Z-strength is essentially a measure of the fiber strength and thus predominantly determined by cellulose. In contrast, f-strength is additionally affected by network effects and thus strongly determined by the cell wall matrix material, i.e. hemicelluloses and lignin (Derbyshire et al. 1995; Xie et al. 2007; Klüppel and Mai 2012). Thermal modification leads to an intensive degradation of hemicelluloses, with this effect being increased for higher temperatures and longer treatment durations (Tjeerdsma et al. 1998; Kotilainen et al. 2000). Consequently, strength loss is higher for the thermal modification at 212 °C compared to the modification at 190 °C.

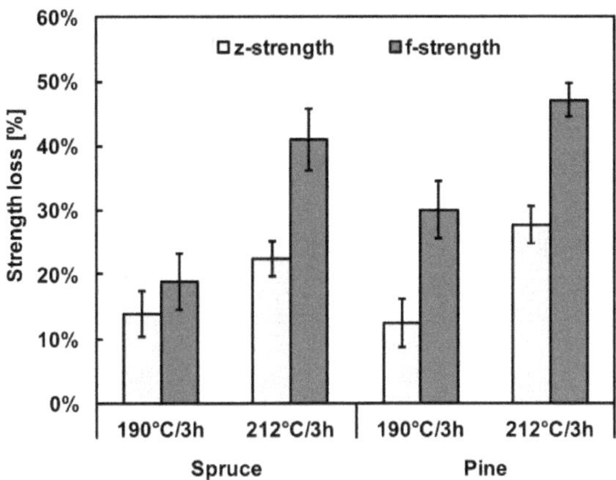

Figure 26: Average micro-tensile strength loss of thermally-modified spruce and pine compared to the respective unmodified references prior to artificial weathering. Strength loss of the respective reference was set to 0 %. (±95% confidence interval)

During artificial weathering, the penetration of the UV-light is expected to be limited to a depth of 75 µm (Hon and Chang 1984), while the micro-veneers have a thickness of 100 µm to avoid defects during handling of thermally-modified veneers. It is thus likely that they are not fully penetrated by the UV-light. UV-light induced degradation, however, can even be found in depths exceeding 100 µm, which can be explained by radical chain reactions that lead to secondary effects migrating deeper into the wood than the UV light itself (Hon and Ifju 1978; Kataoka and Kiguchi 2001). It is therefore expected that the whole thickness of the micro-veneers is affected by photodegradation and that the changes in f- and z-strength can directly be related to photodegradation.

It is evident from **Figure 27** that f- and z-strength decrease with increasing exposure to artificial weathering, caused by photodegradation. This strength loss is found to be almost linear. For all varieties, the loss in f-strength by photodegradation is considerably higher than the loss in z-strength during artificial weathering up to 144 hours. A more rapid degradation in f- than in z-strength of unmodified micro-veneers upon weathering has been reported previously (Derbyshire et al. 1995; Derbyshire et al. 1996). As mentioned above, z-strength is mostly determined by cellulose, while f-strength is strongly affected by hemicelluloses and lignin. It has been reported that besides the predominant degradation reactions within the lignin, carbohydrates are affected by UV-degradation reactions as well (Hon and Chang 1984). However, due to the highly organized structure, crystalline cellulose is hardly susceptible to water, chemicals or UV irradiation, thus limiting the UV-degradation mainly to the amorphous carbohydrates, i.e. hemicelluloses (Hon and Feist 1981).

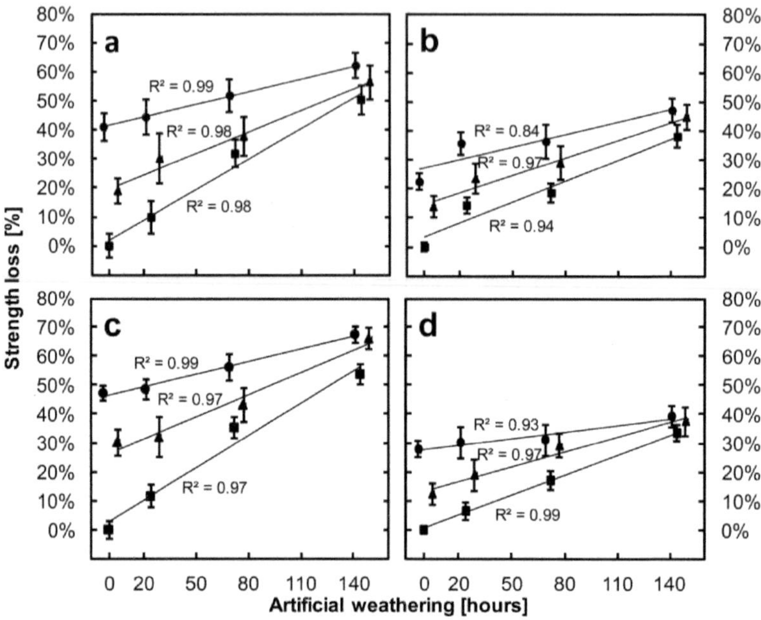

Figure 27: Average micro tensile strength loss [%] during artificial weathering [hours] with respective linear regression curves. a) f-strength of spruce; b) z-strength of spruce; c) f-strength of pine; d) z-strength of pine. Boxes: reference; triangles: 190°C/3h; circles: 212°C/3h. Strength loss of the respective unweathered reference was set to 0 %. Data points were horizontally shifted for better visibility. (±95% confidence interval)

Remarkably, the thermal modification process has a considerable impact on the strength loss by photodegradation, i.e. the loss in f-strength. While a considerable difference is evident between the tensile strength of thermally-modified veneers and the respective reference prior to weathering, the differences in strength diminishes after 144 hours of artificial weathering. Thermal modification decreases the rate at which strength loss by photodegradation occurs. These findings seem to contradict the FT-IR spectroscopic measurements that showed no increased stability of the lignin in thermally modified wood during weathering. However, Klüppel and Mai (2012) reported that a delignification of micro veneers did not influence the f- or z-strength when measuring the veneers in conditioned (20 °C / 65 % rh) state, but rather decreased the tensile strength in wet state. Therefore, degradation of the lignin during artificial weathering is unlikely to be the crucial factor for the loss in strength measured in this study. The decrease in tensile strength, i.e. in f-strength, during artificial weathering can rather be attributed to the degradation of amorphous carbohydrates. As amorphous carbohydrates are already degraded to a high extent in thermally-modified wood prior to weathering, a severe loss in micro tensile strength is evident for non-weathered veneers, but micro tensile strength is less affected by photodegradation than in case of the unmodified references. This coincides with the outcome of measurements by means of FTIR, UV resonance Raman and ^{13}C CPMAS NMR spectroscopy of thermally modified Scots pine exposed to natural weathering that were conducted by Nuopponen et al. (2004). They found fewer changes in the polysaccharide content between weathered and non-weathered samples for thermally-modified pine compared to unmodified pine, because the amorphous carbohydrates were already degraded during thermal modification.

6.4 Conclusions

Chemical alterations of the lignin structure and an increase in the relative percentage of lignin during the thermal modification process of spruce and pine do not prevent photodegradation of lignin. After 144 hours of artificial weathering, the peak at 1508 cm^{-1} disappeared from the FT-IR spectra, irrespective of the thermal modification process. Prior to weathering, loss in micro-tensile strength was evident for thermally-modified wood. However, thermal modification decreased the rate at which strength loss during weathering took place. It is concluded that the photodegradation of amorphous carbohydrates is less pronounced for thermally-modified wood, because the carbohydrates are already degraded during the modification process.

6.5 Acknowledgements

The authors thank the International ThermoWood Association (Helsinki, Finland) for their financial support, the provision of test material and fruitful discussions.

Chapter 7   Publication VI

Thermally-modified Scots pine and Norway spruce wood as substrate for coating systems

Michael Altgen* und Holger Militz

Wood Biology and Wood Products, Burckhardt Institute, Georg-August University Göttingen, Büsgenweg 4, 37077 Göttingen, Germany

corresponding author:
Phone: +49-551 3933664
Fax: +49 551 399646
E-mail: maltgen@gwdg.de

Originally submitted to:
Journal of Coatings Technology and Research
Springer-Verlag Berlin Heidelberg
ISSN (Print)   1547-0091

Submitted: 13 June 2016 / Accepted: 20 August 2016

Abstract: Thermally modified wood (TMW) is increasingly used in exterior applications as alternative to tropical hardwoods or wood impregnated with biocides. Despite its enhanced biological durability and dimensional stability, a surface treatment of TMW with coating systems can be required in certain applications. This study assessed material characteristics of Norway spruce and Scots pine wood that was thermally modified according to the ThermoWood® process and their effect on the performance of commercially available coating systems: a solvent borne oil, a water-borne alkyd reinforced acrylate paint and a water-borne acrylate paint. Residual extractives and remaining degradation products were found in TMW, which carry the risk of causing discoloration or of interfering with the curing reactions of coating systems. The penetration of coating systems into TMW was not found to differ from unmodified wood, although an excessive penetration of solvent-borne oil was found occasionally for TMW. The adhesion strength of water-borne coatings depended on the system that was used. While one system performed sufficiently on TMW, the other coating systems showed a considerable reduction in adhesion strength already after a mild treatment (< 200°C). This reduction could not be assigned to the increase in hydrophobicity of TMW that was evident from contact angle measurements, but was rather related to the mechanical interaction of the specific substrate/coating system.

Keywords: adhesion strength, material characteristics, coating penetration, solvent-borne oil, water-borne coating systems

7.1 Introduction

The understanding of the interaction between coating systems and thermally modified wood (TMW) as substrate is essential for an optimization of the performance in exterior applications. TMW is increasingly recognized as environmental-friendly alternative to tropical hardwoods or wood impregnated with biocides. It is used in many exterior applications, such as decking, cladding or joinery, and various industrially implemented processes exist in Europe (Militz and Altgen 2014). Thermal modification (TM) processes are based on exposing wood to elevated temperatures while minimizing the oxygen-content in the surrounding atmosphere, e.g. by steam, inert gas or oil. One of the predominant process technologies on the European market is the ThermoWood® process which uses superheated steam at atmospheric pressure and applies a high-temperature drying step (100-130 °C) to decrease the wood moisture content severely before raising the temperature up to 200 or 212 °C for hardwoods or softwoods, respectively (Mayes and Oksanen 2003).

TMW is particularly suited for exterior applications due to its improved dimensional stability (Popper et al. 2005) and biological durability (Kamdem et al. 2002). However, its reduced mechanical strength and ductility (Boonstra et al. 2007a) can be a limiting factor. This alteration of technological-relevant properties is primarily caused by chemical changes of the wood composition during the TM (Tjeerdsma et al. 1998). These chemical changes of the wood start with the cleavage of acetyl-groups of hemicelluloses resulting in the formation of acetic acid (Tjeerdsma and Militz 2005). In addition to acetic acid, formic acid is formed from the carboxylic group of the pentosan-glucoronic chain of the hemicelluloses (Bourgois and Guyonnet 1988). These organic acids catalyze the subsequent thermal degradation of amorphous carbohydrates (Tjeerdsma and Militz 2005; Wienhaus 1999; Bourgois and Guyonnet 1988). With the lignin being more temperature-stable, its relative amount increases upon TM and cross-linking reactions take place (Sivonen et al. 2002; Tjeerdsma et al. 1998). However, these chemical changes do not protect the wood from discoloration when exposed to outdoor conditions, thus the original brown color turns grey if no coating is applied (Jämsä et al. 2000; Huang et al. 2012). Furthermore, cracking of TMW in outdoor conditions or during repeated wetting and re-drying is not improved compared to unmodified wood (Feist and Sell 1987; Jämsä et al. 2000; Altgen et al. 2015). Therefore, a surface treatment of TMW with coating systems can be needed depending on the requirements of the application and the end-user.

When using TMW as a substrate, the improved dimensional stability should reduce stresses within the coating caused by movements of the substrate and thus prolong the service life of coatings in exterior applications (Podgorski and Roux 1999). Furthermore, TM leads to a removal of native wood extractives (Poncsak et al. 2009) which can cause discolorations of coatings i.e. by staining from extractives in the knot area (Nussbaum 2004). However, studies on the performance of coating systems on TMW show varying results, since they are based on different TM technologies, wood species or coating systems. On the one hand, there are studies showing that TMW is comparable to unmodified wood as a substrate for coating and that no alterations in the coating recommendation are required (Jämsä et al. 2000; Grüll et al. 2010). On the other hand, some studies report on poor performances of selected coating systems on TMW. Feist and Sell (1987) found that a film forming finish (commercial alkyd resin product) performed worse on thermally modified Norway spruce (*Picea abies* (L.) Karst.) wood than on the reference with respect to the formation of cracks and changes to the finish appearance during natural weathering. After thermal modification of several wood species at peak temperatures above 180 °C, Kesik and Akyildiz (2015) report on a decrease in adhesion strength of a two components water-borne coating. A similar finding was made by de Moura et al. (2013) for a UV-curable polyurethane coating and thermally modified eucalypt (*Eucalyptus grandis*) and pine (*Pinus caribaea* var. *hondurensis*) wood as substrate. Furthermore, coating cracks and adhesive coating failures occurred after artificial weathering of coated TMW in contrast to the coated references. These findings might be related to changes in the material characteristics during the TM process of wood that affect the performance of coatings, which are intended for the application on unmodified wood.

For a good performance of a coating in exterior applications, an effective adhesion is a prerequisite. As comprehensively reviewed (Pizzi 1992; Allen 1987), adhesion can be described by different phenomena, such as mechanical interlocking, diffusion mechanism, electrostatic forces, adsorption by secondary forces or covalent chemical bonding. Although an effective mechanical interlocking is a beneficial effect, the adhesion by secondary forces is regarded as the pre-dominant component in wood adhesion (Pizzi 1992). By thermal modification, wood becomes more hydrophobic, which affects water sorption (Kollmann and Schneider 1963), capillary water uptake (Metsa-Kortelainen et al. 2006), or the wettability (Metsä-Kortelainen and Viitanen 2012; Gérardin et al. 2007; Petrič et al. 2007). In addition to the increased hydrophobicity, the wood becomes more acidic after the TM process, as a result of the formation of carboxylic acids, i.e. acetic and formic acid (Sundqvist et al. 2006; Hofmann et al. 2013). When applying coating systems to TMW, this decrease in pH might affect the curing reactions of the coating. Finally, wood anatomical changes such as damages to the ray parenchyma cells and to the epithelial cells around resin canals or pit deaspiration might create a more open structure for some softwood species and thus potentially affects the penetration of coating systems (Boonstra et al. 2006a; Awoyemi and Jones 2011).

This study assessed if changes of TMW as a substrate affect the performance of coating systems. First, changes in the amount of soluble extractives, acidity and contact angle during the TM were assessed with regard to their potential impact on the wood-coating interaction. Subsequently, the penetration and the adhesion strength of selected coating systems that are typically used on unmodified wood were investigated on TMW surfaces.

7.2 Material and methods

7.2.1 Material

Pre-dried boards of Scots pine (*Pinus sylvestris* L.) and Norway spruce (*Picea abies* (L.) Karst.), with a length of 2 m and dimensions specified in **Table 10** and typically used for the product classes decking, cladding or joinery, were thermally modified according to the ThermoWood® process (Mayes and Oksanen 2003). Two treatment schedules, one resulting in a very mild treatment, the other one in a severe treatment, were applied for each product class in a laboratory treatment reactor using the peak temperatures and durations given in **Table 10**. Besides a peak temperature of 212 °C that is applied in the standard ThermoWood® class "Thermo-D", a peak temperature of 190 °C or lower was studied additionally. Unmodified reference material originating from the same log as the treated boards was available for each product class.

Table 10: Board dimensions and process conditions applied during the ThermoWood® process.

Product class	Wood species	Board dimensions (mm)	Peak temperature and duration	
			Mild treatment	Severe treatment
Decking	Scots pine	32×125	190°C for 3h	212°C for 3h
	Norway spruce	32×150		
Cladding	Scots pine	25×125	180°C for 2h	212°C for 3h
	Norway spruce	25×150		
Joinery	Scots pine	50×125	180°C for 3h	212°C for 4h
	Norway spruce	50×150		

7.2.2 Material characteristics

Extractive content: For the quantification of the amount of soluble extracts, material was collected from ten boards per wood species (pine and spruce) and treatment level (reference, mild and severe treatment), originating from the cladding material (see **Table 10**). The material was either collected from clear wood, with a minimum distance of 50 mm from any knot, or from the knot area, using a drill. For each variety, material from all ten boards was homogenized by milling and mixing in a cutting mill (SM2000, Retsch, Haan, Germany) with a mesh size of 2 mm. The extraction was performed using a Soxhlet apparatus with 6 g of dry wood particles and 200 ml of deionized water as well as ethanol-cyclohexane (1:2, v:v) as a solvent. After extraction, the solution of solvent and extractives was separated using a vacuum-rotary evaporator. The extractive-content was calculated in percentage of the dry mass of the wood material (%). Measurements were done in duplicate for each variety, with the two measurements deviating by less than 8 % from the respective mean value.

Acidity: The same material used for Soxhlet extractions was also used to measure the buffering capacity. 25 g dry wood particles and 300 ml deionized water were added into 1000 ml beakers and placed on a flatbed horizontal shaker for 24 hours. Extracts were filtrated and washed four times to a final volume of 1000 ml. Acidity was measured by titration using 200 ml of water extract and 0.025 M sodium hydroxide. The amount of sodium hydroxide required to reach the neutralization point was determined from the titration curve. This amount was used as a measure for the acidity of the wood extracts in mmol NaOH equivalents per 100 g dry wood (mmol $100g^{-1}$). The measurements were done in duplicate for each variety, with the two measurements deviating by less than 5 % from the respective mean value.

Contact angle of water: The contact angle of deionized water was measured on samples with dimensions of 40×15×100 mm^3 (T × R × L); growth rings were oriented 45° with the tangential surface. Planed surfaces were prepared from four boards per variety, originating from the cladding material (see **Table 10**). Prior to the sample preparation, the boards were conditioned at 20 °C and 65 % relative humidity until the weight change was <0.1 % (w/w) $24 h^{-1}$. The samples were taken from the longitudinal edge (near the bark) of the boards and thus consisted mainly of sapwood. Within 24 hours after the sample preparation the contact angle of deionized water was measured following the sessile drop technique by using a Krüss G10 measurement system in connection with the corresponding Krüss DSA 1 software (Krüss GmbH, Hamburg, Germany). After applying a volume of 10 µl to the wood surface, the contact angle was recorded with 25 frames per second and a total of 250 frames. A minimum of 20 measurements per wood species (pine and spruce) and treatment level (unmodified, mild and severe treatment) was conducted. The contact angle was determined as constant wetting rate angle (CWRA) using the differential method described by Nussbaum (1999). Besides clear wood (at least 50 mm distance from any knot), additional 10 measurements per wood species and treatment level were performed within the knot area. The measurement of clear wood was repeated after storing the samples for 15 days at a constant climate of 20 °C and 65 % relative humidity. Statistical analysis of the TM effect on the CWRA was performed with the Origin 8G system. One-way ANOVA and Tukey's pair wise comparison at a 95 % confidence interval were used to investigate differences between TMW and the reference. Non-significant differences are indicated by a p-value higher than 0.05.

7.2.3 Coating performance

Coating application: For the tests involving coating systems, samples with dimensions of 40×20×300 mm³ (T × R × L) were used; growth rings were oriented 45° with the tangential surface. The samples were prepared using different surfacing techniques: (1) planing, (2) sanding with 100 grit and (3) sanding with 40 grit sanding paper. After sanding, the surfaces were cleaned using pressurized air. All samples were cut from conditioned boards and coated within 24 hours after the sample preparation. For each product class a commercially available coating system was selected that is originally intended for the use on unmodified wood in the respective product class. For the decking material a non-film forming (penetrating) solvent-borne oil (coating A) was applied by brush in a first coat with 100 g m⁻² followed by a second coat with 80 g m⁻². For the cladding and joinery material, a water-borne alkyd reinforced acrylate paint (coating B) and a water-borne acrylate paint (coating C) was selected, respectively. For both product classes, the coating system was applied by spray with a nozzle tip of 1.8 mm and a pressure between 0.2 and 0.3 MPa. The same priming oil was applied as a first coat, before applying the respective coating system in two layers (coating B or C). The solid content and the spreading amounts of the coating systems are given in **Table 11**.

Table 11: Solid content and recommended spreading amounts of the coating systems used for the different product classes.

Code	Product	Description	Solid content (%)	Spreading amounts (unthinned)	
				Wet/one layer (g m⁻²)	Theoretical dry film thickness (μm)
A	Decking	Non-film forming solvent borne oil	25	-	-
Cladding and joinery		Water-borne priming oil	10 (v/v) 11 (w/w)	80	10
B	Cladding	Water-borne, alkyd-reinforced acrylate paint	35 (v/v) 47 (w/w)	135	30
C	Joinery	Water-borne acrylate paint	42 (v/v) 53 (w/w)	155	50

Microscopic observations: Microscopic investigations of the penetration of the respective coating systems into the wood structure and different anatomical features at the wood surface were performed for all product classes. For the decking material with coating A, small blocks (approx. 10×10×10 mm³) were taken from the center of each test piece and were vacuum-impregnated with deionized water (30 min. at 13 kPa). Sections (25-40 μm) were prepared using a sliding microtome (Sartorius Type 31A30, Göttingen, Germany), stained with 1 % safranin and mounted on glass slides, before viewed under an Eclipse E600 fluorescence microscope with a DXM 1200 digital camera (both Nikon, Düsseldorf, Germany) using a G-2A filter. For the cladding and joinery material, thin blocks (approx. 5 mm in thickness) were collected from the central part of each sample and the transverse or longitudinal surface was smoothened using a sliding microtome. The smoothened surface was viewed under the fluorescence microscope using a UV-2A filter as well as under a reflected light microscope (Axioplan 2 Imaging, Zeiss, Oberkochen, Germany).

Adhesion: Adhesion strength was assessed for the cladding and joinery material that were coated with water-borne coating systems using the pull-off test on the basis of EN ISO 24624 (2003) as well as the cross-cut test on the basis of EN ISO 2409 (1994). Only samples with planed surfaces were used for the analysis of adhesion strength. For the pull-off test, dowels with a diameter of 20 mm were bonded to the coating using a two-component adhesive (Araldite 2011, Huntsman Advanced Materials, Everberg, Germany). The coating surrounding each dowel was carefully removed 24 hours after the adhesion of the dowels. Using the PosiTest pull-off adhesion tester, the dowels were detached from the surface in a direction perpendicular to the substrate and the required force was recorded. A minimum of 12 dowels per wood species (pine and spruce) and treatment level (reference, mild and severe treatment) was measured. For the cross cut test, a right angle lattice was cut into the coating with a sharp blade and a distance of 2 mm between the cuts. The cuts were done at 45° to the direction of the grain. A transparent pressure sensitive tape was attached to the lattice while ensuring good contact to the coating. The tape was then pulled off steadily in an angle of approximately 60°. The cross-cut area was examined and classified from 0 (very good adhesion) to 5 (poor adhesion) based on the amount of flaked coating. A minimum of four cross-cuts was tested for each variety.

7.3 Results and discussions

7.3.1 Characteristics of thermally modified wood as a substrate

The wood material as a substrate was characterized before applying coating systems in order to identify changes caused by the TM that might affect the coating performance. **Table 12** summarizes the determination of extractive content and acidity. For clear wood, the extractive content did not decrease but remained either constant (ethanol-cyclohexane-soluble extractives) or even increased (water-soluble extractives) with increasing treatment intensity. The extractive amount determined within the knot area exceeded the amounts in clear wood, with the by far highest amounts measured with ethanol-cyclohexane for the knot area of unmodified pine. TM reduced the amount of extractives in the knot area, i.e. in case of ethanol-cyclohexane soluble compounds. However, even after TM, the amount of extractives in the knot area still exceeded the amount determined for clear wood. The results strongly indicate that native extractives, i.e. non-polar extractives in the knot area, become increasingly volatile at elevated temperatures and are emitted from the wood during the TM process. On the other hand, new, mainly polar extractives are formed as a consequence of thermal degradation of wood compounds that partly accumulate within the wood. The results thus coincide with the findings of Poncsak et al. (2009). Residual native extractives as well as newly formed degradation products might thus still carry the risk of causing discolorations for coating systems applied to TMW. This is particularly critical since many degradation products found in TMW, such as aldehydes or phenolic compounds (Karlsson et al. 2012; Hofmann et al. 2013), contain conjugated double bonds and therefore act as chromophores.

Table 12: Amount of soluble extractives (%) and acidity (mmol 100 g^{-1}) listed for the reference (Ref) as well as for mild and severe treatments of Norway spruce and Scots pine with material collected from regular wood and the knot area.

| Species | Region | Treatment | Extractives (%) | | Acidity (mmol 100 g^{-1}) |
			Water	Ethanol-cyclohexane	
Spruce	Clear wood	Ref	1.02	1.23	0.62
		mild	2.30	1.39	2.34
		severe	2.65	1.33	1.89
	Knot area	Ref	6.74	5.34	0.86
		mild	5.44	3.38	2.90
		severe	5.93	3.20	2.37
Pine	Clear wood	Ref - Sapwood	2.78	3.37	1.45
		Ref - heartwood	1.81	3.40	1.08
		mild	2.97	3.65	3.58
		severe	3.35	3.10	2.16
	Knot area	Ref	6.30	23.18	1.33
		mild	5.12	16.34	3.58
		severe	5.17	11.65	2.40

The acidity (**Table 12**) that was measured for cold water extracts does not correlate with the amount of soluble extractives determined for the same material. The acidity increased upon TM for both wood species, although pine generally featured a higher acidity than spruce. The increase in the acidity of the wood after the TM can be attributed to the formation of carboxylic acids. Acetic and formic acids are the predominant acids that are produced during the TM of wood. The formation of acetic acid is caused by the cleavage of acetyl groups of hemicelluloses (Tjeerdsma et al. 1998; Manninen et al. 2002; Sundqvist et al. 2006), which already starts at temperatures lower than 200 °C (Wienhaus 1999). Accordingly, a high acidity is already evident after a mild treatment (180 °C for 2 h). A severe treatment (212 °C for 3 h), however, lead to a lower acidity with 1.89 for spruce and 2.16 mmol 100 g^{-1} for pine. Although carboxylic acids are still being formed during the TM process, they are vaporized and emitted from the wood and the treatment kiln at higher temperatures (Hofmann et al. 2013; Altgen et al. 2014).

Acidic extractives in some wood species interfere with the curing and hardening reactions of adhesive and coating systems and are associated with increased brittleness of the coating film, reduced adhesion strength, and peeling of the coating from the wood (Stefke and Dunky 2006; Wellons et al. 1977; Hse and Kuo 1988). A similar effect might be caused by carboxylic acids that remain within the wood after TM. The low pH of Norway spruce after

TM according to the PLATO® process has been previously linked to insufficient hardening and poor adhesion of a phenol-resorcinol-formaldehyde adhesive (Sernek et al. 2008). Vice versa, the preferential removal of acidic extractives during the TM of European Ash (*Fraxinus excelsior* L.) decreased the acidity and was considered as the cause for an increased adhesion of a water-borne coating that included an alkyd binder in a recent study by Herrera et al. (2015).

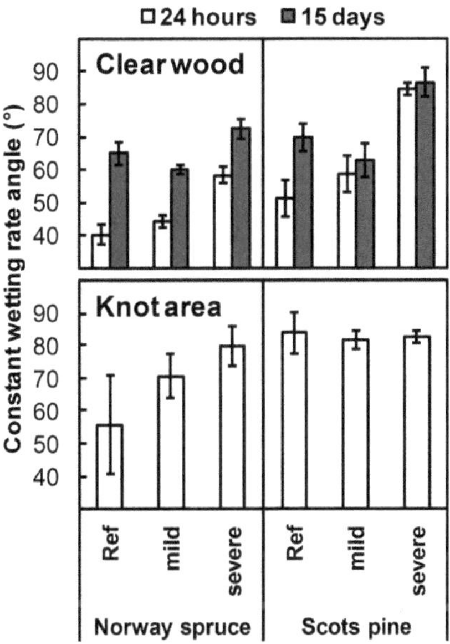

Figure 28: Average constant wetting rate angle of the reference (Ref) as well as mild and severe treatments of Norway spruce and Scots pine. Top: measurement on clear wood 24 hours and 15 days after planing; Bottom: measurement within the knot area 24 hours after planing. (±95% confidence interval)

When a water droplet is placed on a wood surface for contact angle measurement, two main processes occur as a function of time: spreading onto the surface and penetration into the wood bulk. In contrast to dynamic contact angle measurements, these two processes cause a non-linear change in contact angle in dependence on the time after droplet application when using the sessile drop technique. The determination of the CWRA considers the different processes by dividing the change in contact angle as a function of time in a spreading stage with a rapid change in contact angle, and a penetration stage with a slower and nearly constant change in contact angle (Nussbaum 1999). This is not considered when simply measuring the contact angle after a specific time.

The CWRA, shown in **Figure 28** A and calculated for contact angle measurements with water on planed surfaces, is influenced by the TM process as well as by the storage duration. For measurements within 24 hours after the sample preparation, the unmodified reference material featured the lowest CWRA with 40.41° for spruce and 51.33° for pine. TM increased the CWRA, reaching 58.45° and 84.75° after the severe treatment of spruce and pine, respectively. However, a mild treatment did not change the CWRA significantly for spruce or pine (p = 0.33 and 0.16). An increase in the contact angle of water after TM was recorded in several studies (Hakkou et al. 2005; Gérardin et al. 2007; Metsä-Kortelainen and Viitanen 2012), but the results are not completely consistent, potentially due to differences in the measurement technique and the storage duration that was applied as well as due to differences in the TM technology and variations in the wood material. Increased contact angles of water have previously been explained by plasticization and reorganization of the lignocellulosic compounds (Hakkou et al. 2005) as well as by the decrease of free reactive hydroxyl groups in TMW (Gérardin et al. 2007).

Repetitive measurements after storing the samples at a constant climate for 15 days resulted in an increase of the CWRA. Such an ageing effect was less pronounced for TMW compared to the unmodified references. For unmodified spruce and pine the CWRA increased to 64.95 and 69.73°, respectively, which was even higher than the CWRA measured for material that was thermally modified by a mild treatment (60.08 and 62.91°). The CWRA of pine that was thermally modified at high intensity remained almost unchanged after storage. Generally, 15 days of storage diminished differences in the CWRA between the varieties compared to a measurement on freshly prepared samples. An increase in the contact angle of water during storage of unmodified wood is a well known phenomenon (Nussbaum 1999; Gindl et al. 2004a; Gardner et al. 1991; Nguyen and Johns 1979; Wålinder 2002). It can be explained by the migration of extractives from the interior to the exterior of the wood, which creates a hydrophobic surface, as well as by the reorientation of functional groups at the wood-air interface (Nussbaum 1999; Wålinder 2002). XPS measurements on aged wood surfaces indicate a decrease in polarity due to the decrease of oxygen and the increase of carbon percentage at the surface, which points towards a higher hydrophobicity (Gardner et al. 1991; Gindl et al. 2004a). This decrease in polarity upon ageing appears to be less pronounced on TMW surfaces which already feature reduced oxygen and increased carbon percentages prior to ageing (Gérardin et al. 2007; Inari et al. 2006).

The CWRA measured within the knot area (**Figure 28** B), generally exceeds the CWRA measured on clear wood and suffers from high data variation in case of unmodified spruce. Although differences in surface roughness or grain angle complicate the comparison of clear wood and the knot area, the increased CWRA on the knot area coincides with the higher amount of ethanol-cyclohexane soluble extractives. In case of unmodified pine, which featured the highest amount ethanol-cyclohexane soluble extractives in the knot area, the CWRA was considerably higher than for clear wood. Consequently, no significant differences in the CWRA between thermally modified and unmodified pine can be observed for measurements within knot area. Therefore, extractives within the knot area as well as an increase in the storage duration diminish an impact of the TM process on the CWRA.

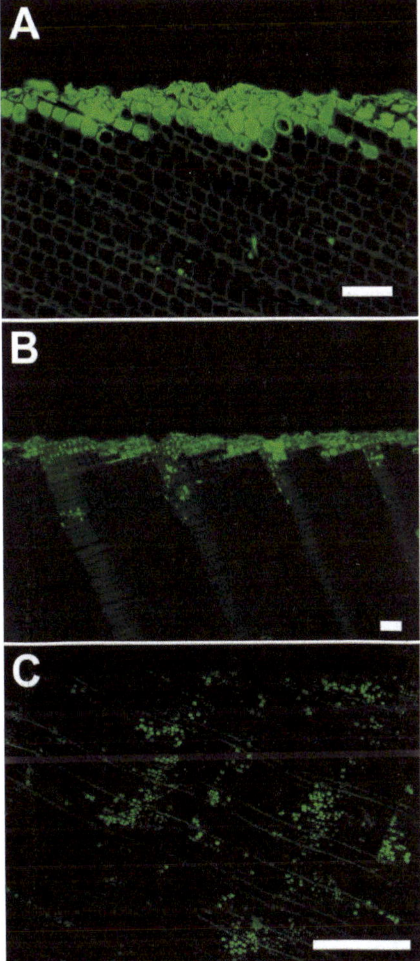

Figure 29: Transverse sections showing the penetration of the solvent-borne oil investigated by fluorescence microscopy and a G-2A filter. A and B) unmodified Scots pine sanded with 100 grit sanded paper (scale bars 100 µm), C) Thermally modified Scots pine with planed surface showing an area with high penetration (scale bar 1000 µm).

7.3.2 Microscopic observations

The penetration into the wood substrate mainly depended on the coating system that was applied and was to a much lesser extent influenced by the surface preparation technique or the TM process. The solvent borne oil (coating A) filled the outer tracheids and the second cell row, sometimes even the third (see **Figure 29** A). Lumens filled with coating A could also be found in some distance from the surface, especially in the latewood region (see **Figure 29**

B) and more often in case of pine than for spruce. This deep penetration is most likely linked to the penetration through the rays and the spread of coating into adjacent longitudinal tracheids, as described by Nussbaum (1994) and de Meijer et al. (1998) for solvent-borne alkyd paints. The spread into longitudinal tracheids might be facilitated in the latewood, because the aspiration of bordered pits is less complete in the latewood tracheids than in the earlywood tracheids and because of the insertion of parenchyma cells in the ray tracheid rows at the boundary of an annual ring (Nussbaum 1994). Furthermore, capillary forces that drive the penetration of the solvent borne oil are higher in the smaller lumen of the latewood tracheids. In contrast to the solvent-borne oil, the penetration of the two water-borne coating systems applied to the cladding (coating B) and joinery material (coating C) was limited to the flow into outer tracheids that were cut open by the surface preparation (**Figure 30** A). Occasionally, coating could be found one cell row lower, which can be explained by the flow of paint into the open end of a tracheid over a short distance (**Figure 30** B), as shown by de Meijer et al. (1998). There was, however, no sign of a penetration from cell to cell via the pits or through the ray cells.

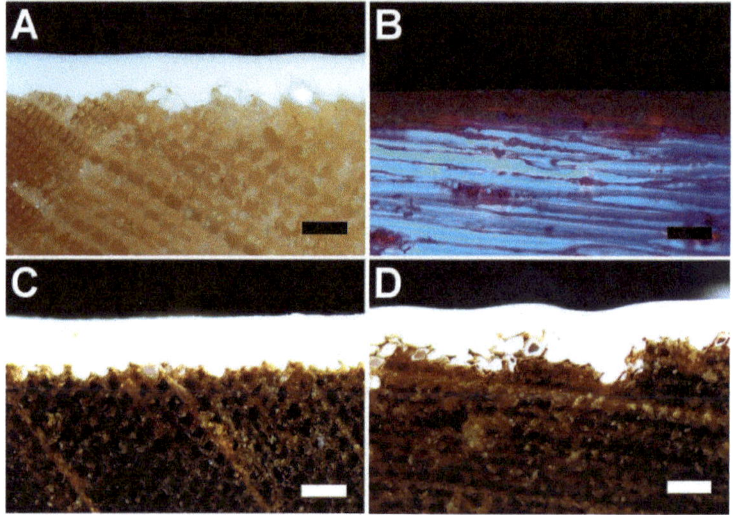

Figure 30: Penetration of water borne coating systems. A) Transverse section of unmodified pine with planed surface and coating B; B) Oblique tangential section of unmodified pine with sanded (100 grit) surface and coating B, showing the flow of coating into open tracheid ends (fluorescence microscopic image using a UV-2A filter); C) Transverse section of spruce treated at 212°C/4h with planed surface and coating C; D) transverse section of pine treated at 212°C/3h with sanded surface (40 grit) and coating B. Scale bars: 100 µm

On TMW, the penetration of the water-borne coating systems was not found to differ from the penetration on unmodified pine and spruce that is described above (**Figure 30** C). Additional features that might indicate a weak boundary layer, such as air bubbles at the interface, which might be created if the coating cures prior to a complete wetting, were not observed for the water-borne coating systems applied to TMW surfaces. In case of the solvent-borne oil, large sections with filled lumen of longitudinal tracheids were found occasionally in depths exceeding 1000 µm for pine and spruce that were modified in a severe treatment (**Figure 29** C). This feature was found independent of the wood species or the surface preparation technique (planing/sanding). As a potential explanation, microscopic defects such as the destruction of ray cells that has been reported for thermally modified softwoods (Boonstra et al. 2006a; Awoyemi and Jones 2011; Altgen et al. 2015), might facilitate the above described penetration through the rays and the spread of the coating into the adjacent longitudinal tracheids. TMW might therefore require a higher wood oil coverage per area than unmodified wood to ensure a sufficient amount of oil at the surface layer where it is required.

As to be expected, the main impact of the surface preparation technique was a change in the structure of the outer tracheids of pine and spruce. In line with the investigations of de Meijer et al. (1998), planed surfaces showed less evidence of compressed cells or damaged cell walls than sanded surfaces. In especially, the latewood tracheids appeared almost entirely intact for planed surfaces. Sanding resulted in compression of earlywood tracheids and a higher extent of damages to the outer cells, as can be observed in **Figure 29** A. This effect was of course most pronounced for a very coarse sanding paper (40 grit). The very rough surface of samples sanded with a coarse sanding paper also resulted in an uneven thickness of the two film-forming, water-borne coating systems with partly detached fibers occasionally reaching through the coating film. By creating more damage to the outer cells, the sanding process led to a higher surface area. Although this did not affect the penetration depth, it affected the amount of solvent based wood oil that was taken up. Consequently, planed surfaces featured residual oil at the surface, while this could not be observed for sanded surfaces. For the water-borne coating systems, a higher surface area might be beneficial in terms of adhesion by secondary forces on the one hand, because the damages to the outer cells result in more open connection for the coating to flow into. On the other hand this might also arise in a mechanical weak surface layer, as described by Stehr and Johansson (2000).

7.3.3 Adhesion strength

For the two water-borne coating systems, the adhesion was assessed using the pull-off test and the cross-cut test, with the two test procedures showing the same basic tendencies (see **Figure 31**). On unmodified wood surfaces, both water-borne coating systems performed well, with pull-off strengths exceeding 3 N mm^{-2} and cross cut test ratings of 1.5 or lower. On TMW surfaces, the pull-off strength depended strongly on the coating system that was applied. Coating C that was applied to the joinery material resulted in similar pull-off strengths on TMW as on the unmodified references. No change in pull-off strength was found in case of thermally modified pine and only a slight reduction to 2.5 and 2.7 N mm^{-2} was evident for spruce that was modified in a mild and severe treatment, respectively. In contrast, coating B that was applied to the cladding material resulted in a considerable decrease in the pull-off strength for thermally modified pine and spruce with a maximum of

2 N mm^{-2}. This decrease was confirmed by the cross-cut test, which resulted in an almost complete flaking of coating B when applied to TMW surfaces, leading to an average test rating of 4.9 or higher.

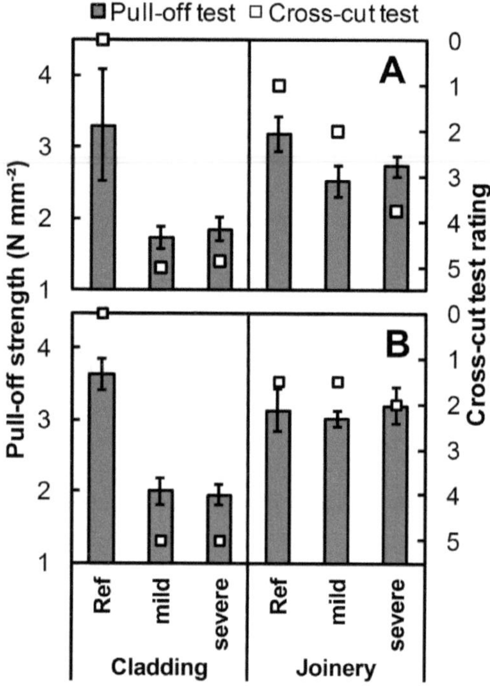

Figure 31: Pull-off strength (N mm^{-2}) and cross-cut test rating for the reference (Ref) as well as for mild and severe treatment intensities of the joinery and cladding material. A) Norway spruce; B) Scots pine. Error bars represent 95% confidence interval.

A reduction in mechanical interlocking or in adhesion by secondary forces would explain the loss in pull-off strength observed for coating B on TMW surfaces. However the investigation of the penetration of the two water-borne coating systems into the wood did not indicate a reduced mechanical interlocking or a reduced area for coating contact for TMW as substrate. Moreover, results of the contact angle of water do not coincide with the measured pull-off strength. A mild treatment with no significant change in the CWRA compared to the reference already resulted in low pull-off strengths for coating B, while a drastic increase in the CWRA after a severe treatment did not reduce the pull-off strength of coating C. Gindl et al. (2004a) stated that the wettability of wood by water does not sufficiently explain the interaction between wood and coating systems. Indeed, Petrič et al. (2007) found much better wetting of water-borne coating systems on TMW than on unmodified wood surfaces, even though the contact angle of water increased upon TM. The hydrophobic character of TMW is therefore considered as insignificant with regard to the adhesion strength measured in the present study and should not be regarded as general drawback for the application of water-borne coating systems.

During the pull-off test, the amount of cohesive failure of wood increased for TMW compared to the unmodified reference. This indicates that the pull-off test was additionally influenced by the reduced strength of TMW as a substrate rather than being solely determined by the interface bonding. A similar observation was made by de Moura et al. (2013) for *Eucalyptus grandis* and *Pinus caribaea* wood samples that were thermally modified at maximum temperatures between 140 and 200 °C and roll-coated with an UV-curable polyurethane coating. They related a reduction in the adhesion strength to a decrease in the mechanical properties during the TM process. However, differences in the pull-off strength and the cross-cut test ratings between the two water-borne coating systems evidence that the adhesion of coatings to TMW is not a simple function of the loss in mechanical strength of the substrate. While a pull-off strength of 2.5 N mm^{-2} or more was achieved for coating C even after a severe treatment, a severe reduction in pull-off strength for coating B was already evident after a mild treatment at which strength loss of the wood is minor (Boonstra et al. 2007a). This leads to the conclusion that the observed loss in adhesion strength, rather, depends on the specific substrate/coating system.

If the interface bonding is good, a brittle coating on a brittle substrate results in a system that tends to behaves in the same way as a brittle bulk material (Bull and Berasetegui 2006), thus leading to a low pull-off strength and poor cross-cut test rating. Under tension, such a system typically leads to film cracking followed by the extension of the cracks into the substrate (Evans et al. 1988). TM of wood results in a significant increase in brittleness even at mild treatments and temperatures as low as 160 °C (Phuong et al. 2007). Due to their three dimensional structure when cured, alkyd resins are more rigid and brittle compared to chain-like acrylic resins. The alkyd reinforcement in coating B (see **Table 11**) might thus lead to a more brittle behavior compared to coating C, which explains the poor adhesion strength of the TMW/coating B system. In this context, it would be interesting to test if the increased acidity of TMW contributed to such an increased coating brittleness. Using a more ductile substrate (unmodified spruce or pine) and/or a more ductile coating system (coating C) seems to improve the adhesion strength assessed by the pull-off and cross-cut test.

7.4   Conclusions

TM of spruce and pine leads to changes that should be considered when applying coating systems to TMW. High amounts of extractable compounds that might cause discolorations, increased acidity due to the formation of carboxylic acids that might interfere with the curing of coatings and an increased contact angle of water after TM and/or storage were recorded for thermally modified spruce and pine. The performance of coating systems strongly depended on the individual substrate/coating. The penetration into TMW was generally not found to differ from unmodified wood, although extensively high penetrations of a solvent-borne oil were occasionally found for thermally modified wood, which might necessitate a higher wood oil coverage per area. Results of the adhesion strength testing suggest that sufficient adhesion strength can be achieved on TMW with water-borne coating systems that are intended for the use on unmodified wood, but it should be verified that the respective coating system can cope with the changes of the substrate induced by TM.

7.5 Acknowledgements

The authors thank the International ThermoWood Association (Helsinki, Finland) for their financial support, the provision of test material and fruitful discussions.

Chapter 8   General discussion

The following chapter connects the individual discussions in **publications I – VI**. It is the aim to go beyond a simple summary of the previous chapters in order to refer to the objectives of this thesis stated in **1.3**. The chapter is divided into two parts. The first part (**8.1**) deals with thermal modification processes in closed reactor systems at elevated pressure and their impact on the thermal degradation and the change in wood properties. The second part (**8.2**) focuses on the effect of abiotic factors on wood that is thermally modified in open reactor systems.

8.1   Improving wood properties by thermal modification in closed reactor systems (I-III)

Studying the impact of elevated water vapor pressure (WVP) on the resulting wood properties requires a suitable basis for comparison. Keeping all process parameters constant and only changing the WVP (*ceteris paribus*) might be the simplest approach. However, it has been previously demonstrated that increasing the WVP during thermal modification processes accelerates thermal degradation reactions within the wood (Seborg et al. 1953; Burmester 1973; 1975; Borrega and Kärenlampi 2008a; Ding et al. 2011b; Torniainen et al. 2011; Rautkari and Hill 2014), which should obviously affect the resulting wood properties, similar to the effect of increased peak temperatures or durations. The more relevant question is, whether TMW properties are a simple function of the generated level of thermal degradation or further influenced by specific process conditions. To answer this question, this chapter first discusses the impact of WVP on the thermal degradation of wood, with wood mass loss as a marker, and then interprets TMW properties as a function of the mass loss generated.

8.1.1   Thermal degradation at elevated water vapor pressure

To quantify thermal degradation of wood during thermal modification processes, a vast number of "markers" exists, as reviewed by Willems et al. (2015b). The most comprehensive marker that has been applied for decades (see e.g. Stamm et al. 1946; Seborg et al. 1953; Stamm 1956) is the mass loss of the wood during the process (see **1.2.4**). This mass loss is generated by the degradation of cell wall polymers and their conversion into components that become volatile at elevated temperatures and are thus removed from the wood. However, results of **publications I and III** clearly demonstrate that the application of elevated WVP results in an accumulation of degradation products within the wood, which reduces the suitability of mass loss as a marker for thermal degradation. Such an observation was also reported by Obataya et al. (2002). The application of elevated pressure hinders the vaporization of volatile compounds formed upon thermal degradation into the gaseous phase of the treatment reactor. Rapid vaporization by boiling and even evaporation of liquids and organic compounds are prevented or at least reduced if the temperature is sufficiently low and if the total and/or the respective partial pressure is sufficiently high. This phenomenon can be easily observed with the concentration of organic acids in thermally modified wood, such as formic and acetic acid, since they have an atmospheric boiling point as low as ca. 101 and 118°C, respectively. The acidity measured in **publication VI** of Scots pine and Norway spruce increased by thermal modification, due to the formation of organic acids. Increasing the peak temperature above 200°C, however, reduced the acidity compared to temperatures below 200°C, due to enhanced vaporization. Thermal modification in sub-atmospheric pressure even reduces the acidity of wood to levels lower than those measured for unmodified wood (Hofmann et al. 2013). For thermal modification at elevated WVP described in **publication I**, a strong increase in acidity is evident for beech, with the highest levels recorded after treatments at a low peak temperature and a high maximum pressure.

The accumulated degradation products are not distinguished from the remaining, unaffected cell wall polymers when calculating the mass loss on the basis of the dry sample mass before and after the process. This leads to an underestimation of the actual thermal degradation occurring in closed reactor systems compared to the thermal modification in open reactors. When estimating thermal degradation, the mass loss thus needs to be based on the dry and extractive-free mass of the wood before and after the process (ML_c, corrected mass loss). The ML_C is also unaffected by a removal of native extractives, which is unrelated to the depolymerization of cell wall constituents. However, it needs to be noted that even the ML_c does not consider chemical reactions such as condensation or esterification reactions, which do not result in a loss in cell wall substance, but rather lead to a fixation of some degradation products within the wood.

In agreement with previous studies (Seborg et al. 1953; Burmester 1973; 1975; Borrega and Kärenlampi 2008a; Ding et al. 2011b; Torniainen et al. 2011; Rautkari and Hill 2014), results of **publication I and III** demonstrate that wood mass loss is strongly affected by the water vapor pressure. Mass loss can be generated in the time frame of several hours at temperatures as low as 135 °C (see **publication III**), if the water vapor pressure is sufficiently high. For Norway spruce, Borrega and Kärenlampi (2008a) did not record any mass loss after treatments at 120 °C even for durations of 8 h and saturated water vapor pressure. The threshold temperature at which mass loss is generated at elevated water vapor pressure within several hours thus seems to be between 120 and 135 °C. However, Borrega and Kärenlampi (2008a) did not consider the potential presence of accumulated degradation products within the wood, which is why the threshold temperature might be even lower than 120 °C.

In **publication I**, a strong dependence of the corrected mass loss on the maximum pressure applied during the process was found for European beech, mostly irrespective of the peak temperature. A similar observation was reported by Willems et al. (2015a) for Norway spruce. A definite explanation for this pressure-dependence could not be found, but potential factors were discussed in **publication I**, which can be summarized as follows:

(1) Improved heat-transfer in a compressed gas atmosphere (Giebeler 1983) and/or reduced evaporative cooling at elevated water vapor pressure (Willems et al. 2015a), which prolong the exposure of the wood to elevated temperatures.

(2) Accumulation of organic acids in the wood during modification at elevated pressure, which catalyze thermal degradation reactions and thus compensate for lower temperatures (Stamm 1956; Willems et al. 2015a).

(3) Increased equilibrium moisture contents of wood at high treatment temperatures as a result of the elevated water vapor pressure environment (Engelhardt 1979; Lenth and Kamke 2001; Ishikawa et al. 2004). The presence of water not only catalyzes hydrolysis reactions (Garrote et al. 1999), but also lowers the activation energy required for thermal degradation by increasing the softening and mobility of amorphous cell wall polymers (Borrega and Kärenlampi 2008a).

To separate these individual factors it needs to be investigated whether the total pressure or the partial water vapor pressure is decisive for the accelerated thermal degradation in closed reactor systems. Factors (1) and (2) should be mainly related to the total pressure, while factor (3) should be mainly related to the partial water vapor pressure in the reactor. Thermal modification processes during which an increased total pressure is generated with a dry inert gas such as nitrogen might therefore be a promising approach to separate the different factors, but were not realized within the framework of this thesis. Thermal modification processes in a closed system in nitrogen atmosphere by Giebeler (1983) indicate strongly that there is an effect of total pressure on thermal degradation, but the peak temperatures applied were considerably higher (≥180°C) than in **publication I** (≤180°C) and **publication III** (≤155°C). On the other hand, a profound impact of the presence of water on the thermal degradation was evidenced by a much higher corrected mass loss of water-saturated Scots pine wood samples compared to oven-dry samples after thermal modification in saturated water vapor (**publication III**). Unfortunately, the separation of the individual factors is further hindered by their interrelation. Although, the accumulation of organic acids might be strongly depending on the total pressure, their formation is catalyzed by the presence of water (Garrote et al. 1999; 2001; Tjeerdsma and Militz 2005).

In addition to the impact of elevated WVP on wood mass loss as a quantitative marker for thermal degradation, it should be noted that elevated WVP and the consequent increase in wood MC during the process (Engelhardt 1979; Lenth and Kamke 2001; Ishikawa et al. 2004) are likely to affect the equilibrium and the rates of degradation reactions. Water is involved in many thermal degradation reactions, either as a reactant or as a product. According to Chatelier's principle, increasing the amount of water present in the wood should thus accelerate reactions with water as a reactant, e.g. hydrolysis, but decelerate reactions with water as a product, such as dehydration, esterification or condensation reactions. In agreement with this explanation, higher reducing sugar contents, but lower furfural contents were found in Scots pine that was thermally modified in water-saturated state compared to a modification in an oven-dry state (**publication III**). This indicates preferential hydrolysis of hemicelluloses, but reduced dehydration of the sugar fragments into aldehydes in the presence of water. The presence of water might thus contribute to the accumulation of degradation products in wood modified at elevated WVP by limiting the conversion of carbohydrates into volatile compounds (i.e. aldehydes).

The shift in the equilibrium of degradation reactions is also evidenced by investigations of the PLATO® process. In this two step process, strong splitting of acetyl groups, carbohydrate chains or ether bonds within the lignin are mainly found in the first process step performed under "wet" conditions in saturated water vapor ("hydrothermolysis step"). On the other hand, condensation and cross-linking reactions within the lignin or esterification reactions are observed frequently in the second process step performed under "dry" conditions ("curing step") (Tjeerdsma et al. 1998; Tjeerdsma and Militz 2005).

8.1.2 Properties of wood thermally modified at elevated water vapor pressure

The wood properties were analyzed as a function of the mass loss recorded after the thermal modification processes. In contrast to many other studies, the corrected mass loss was applied instead of the dry mass loss. This ensures the comparability of the different process conditions, even though the accumulation of degradation products differs for the temperature and pressure regimes applied. In the framework of this thesis, the analysis was limited to changes in physical wood properties, with a special emphasis on wood sorption.

Sorption

The sorption of thermally modified wood was assessed by the EMC at 20 °C and 65 % RH. **Publications II and III** evidence a profound impact of the WVP and the consequent presence of water within the wood during the process on the resulting sorption of the modified wood. In the following, the outcomes of these publications are brought together to explain mechanisms involved in reducing the sorption of wood by thermal modification.

When plotting the EMC that is measured by conditioning directly after the thermal modification process (intial EMC) against the mass loss, the result is not a linear correlation (**publication II**). The reduction in EMC-% per mass loss-% is much higher at low levels of wood mass loss compared to high levels of mass loss, as illustrated in **Figure 32**. Such saturation effects in EMC reduction were also observed in other studies on thermal modification at atmospheric pressure in superheated water vapor (Viitaniemi et al. 1997; Esteves et al. 2007). They report on a strong reduction in EMC at low mass loss levels, but no further reduction in EMC for dry mass loss levels above 6-8 %. However, no explanation for such saturation is given. In addition to this non-linear behavior, the initial EMC as a function of mass loss was strongly influenced by the process RH applied during the thermal modification (**publication II**). Lowering the process RH further reduced the initial EMC, and resulted in an EMC reduction even if no mass loss was generated. The reduction in initial EMC is thus not solely linked to the loss in cell wall substance, but influenced by further mechanisms. A substantial impact of accumulated thermal degradation products within the wood on the initial EMC was ruled out. This was further evidenced by DVS measurements of extracted and non-extracted particles of thermally modified wood. Additional sorption sites provided by degradation products and their cell wall bulking effect might compensate one another. In agreement with Borrega and Kärenlampi (2010) the impact of the process RH was rather related to wood drying and ultra-structural changes of amorphous polymers during the exposure to elevated temperatures.

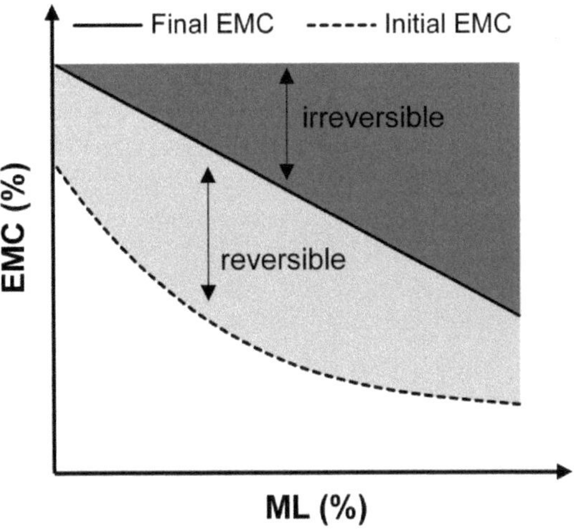

Figure 32: Suggested change in equilibrium moisture content (EMC, in %) in dependence on the mass loss (ML, in %) for wood thermally modified at low process RH. In this scenario, initial and final EMC are both determined after adsorption from the dry state at the same climate. The initial EMC refers to the EMC directly after the modification process, whereas the final EMC refers to the EMC after sufficiently long exposure at nearly saturated water vapor conditions or after water-soaking.

The orientation of cell wall polymers and the cell wall (ultra-) structure in unmodified wood contribute to the hygroscopicity of wood by causing an internal restraint to swelling (Skaar 1988). Ultra-structural changes in wood upon drying that effect the number of available sorption sites were evidenced by Suchy et al. (2010b). They converted accessible OH groups in fresh wood samples to OD groups by deuteration. Without drying, the OD groups were fully reversed back into OH groups after exposure to excess water. However, drying after deuteration converted accessible into inaccessible OD groups, which could not be converted back into OH groups. This was taken as a clear indicator of changes in the wood ultra-structure. The same group of authors suggested that these ultra-structural changes are qualitatively identical to well-known changes occurring in pulp fibers upon drying (Suchy et al. 2010a), which is denoted as hornification (Minor 1994).

However, the studies by Suchy et al. (2010a; 2010b) on the impact of drying on the wood ultra-structure, as well as the hornification effect in pulp fibers (Minor 1994) describe a mechanism that cannot be recovered upon re-wetting. It is often explained by an irreversible aggregation of cellulose microfibrils and the formation of irreversible hydrogen bonds when closing the distance between cell wall polymers during drying (Crawshaw and Cameron 2000; Salmén and Fahlén 2006; Suchy et al. 2010b). Removing the restricting spacers among the fibrils by the degradation of hemicelluloses during the modification process might enhance an irreversible aggregation of cellulose (Duchesne et al. 2001; Fahlén and Salmén 2003; Salmén 2015). Indeed, Andersson et al. (2005) detected an increase in the size of cellulose crystallites for thermal modification of Scots pine wood above 150 °C, which strongly supports this theory. Although such a mechanism is likely to occur during thermal modification, it cannot be easily differentiated from the impact of thermal degradation reactions. Furthermore, it was shown in **publication II** that a substantial proportion of the reduction in initial EMC caused by thermal modification is reversible upon conditioning at high RH or water-soaking, in line with the findings by Obataya and Tomita (2002) or Hill et al. (2012b). The increase in EMC upon water exposure diminished the differences in EMC related to the process RH and resulted in a nearly uniform, linear correlation between the final EMC (measured after conditioning at high RH and water-soaking) and the wood mass loss (see **Figure 32**). A very similar observation was published very recently for Sitka spruce wood that was thermally modified at 120°C and different process RH (Endo et al. 2016).

A possible explanation for the reversible EMC reduction of thermal modified wood is enabled by the consideration of the softening of amorphous wood polymers during the exposure to heat. Thermal modification at low WVP not only results in strong drying, but is also associated with an increase in the wood temperature above the glass transition temperature (T_g), above which the mobility of amorphous polymers is strongly enhanced (Chow and Pickles 1971; Hillis 1984). The increased mobility facilitates ultra-structural realignments of amorphous polymers that might be induced by shrinkage stresses caused by drying and the removal of cell wall polymers at elevated temperatures (Cheng et al. 2007). However, when the temperature falls below the T_g during cooling, the polymer mobility is reduced abruptly and the ultra-structural realignments are locked (Hillis 1984). During conditioning at 20 °C and 65 % RH directly after the process to determine the initial EMC, the amorphous wood polymers are thus restraint from expansion and the cell wall matrix is in a stiffened state. When considering the role of matrix polymer relaxation during wood sorption (Vrentas and Vrentas 1991; Hill et al. 2012a; 2012b; Popescu and Hill 2013), an increase in the cell wall matrix stiffness during thermal modification restricts the expansion of the cell wall nanopores and thereby reduces the availability of water-accessible sorption sites (Hill et al. 2012b). Applying a higher process RH limits excessive drying and the consequent ultra-structural changes, but also increases the wood moisture content during the cooling phase. The presence of moisture reduces the T_g of amorphous cell wall polymers (Chow and Pickles 1971; Salmén and Olsson 1998), and thereby increases the time at which ultra-structural realignments can be recovered. Thus, reversible cell wall matrix stiffening is less pronounced at high process RH.

A recovery of ultra-structural realignments within the cell wall matrix can also be achieved after the thermal modification process, as evidenced by the increase in EMC during sorption cycles or water-soaking. When conditioning wood above 60-75 %, softening of hemicelluloses occurs even at room temperatures. This is linked to the upward bend in the sorption isotherm of wood (Vrentas and Vrentas 1991; Engelund et al. 2013). It enables stress relaxation and the reversal of ultra-structural realignments within the cell wall matrix after the thermal modification process, which decreases the matrix stiffness. Water-soaking of wood has a similar or even stronger effect, because it creates rapid, full water saturation and maximum swelling of the cell wall.

The proposed mechanism of reversible ultra-structural realignments within the cell wall matrix is consistent with the differences in adsorption isotherms of thermally modified wood between the first and subsequent sorption cycles observed by Hill et al. (2012b) and Majka et al. (2016). It does also provide an explanation for the non-linear correlation between initial EMC and mass loss. Ultra-structural realignments of amorphous polymers contribute to a rapid decrease in EMC at low mass loss levels. However, at high mass loss levels large quantities of amorphous polymers, i.e. hemicelluloses, are already removed from the wood. The contribution of ultra-structural realignments of amorphous polymers to the initial EMC reduction is then limited, which results in saturation in the initial EMC reduction (see **Figure 32**).

In agreement with the explanations above, an almost linear decrease of the EMC as a function of mass loss was found in **publication III**, because water-soaking was conducted and drying below fiber saturation did not occur during the modification process in saturated water vapor. Such a linear correlation might indicate that the irreversible EMC reduction by thermal modification is solely caused by the removal of hemicelluloses. Hemicelluloses contain more water-accessible OH groups than cellulose or lignin (Runkel 1954; Runkel and Lüthgens 1956), but are preferentially degraded at elevated temperatures (Alén et al. 2002). However, **publication III** verified an adverse effect of elevated wood moisture contents on the improvement in hygroscopicity and dimensional stability during thermal modification processes, as found in previous studies (Stamm and Hansen 1937; Seborg et al. 1953; Rautkari and Hill 2014). The slope at which the final EMC decreased as a function of mass loss was found to be strongly influenced by the wood moisture content during the thermal modification process. A more efficient EMC reduction was achived by the thermal modification of dry wood compared to the modification of water-saturated wood. Consequently, the irreversible EMC reduction by thermal modification must be influenced by an additional mechanism beyond the removal of accessible OH groups.

A change in the cell wall matrix stiffness is a plausible mechanism for the effects observed in **publication III**. However, in contrast to the above described reversible cell wall matrix stiffening by ultra-structural realignments (**publication II**), the change in matrix stiffness observed in **publication II** must be irreversible as it is still evident after extensive water-soaking. The moisture content during the process might control an irreversible change in cell wall matrix stiffness by its effect on the equilibrium of thermal degradation reactions (see **8.1.1**). The modification of the lignin carbohydrate complex, in particular the enhanced formation of covalent bonds within the cell wall matrix by preferential esterification and condensation reactions under dry conditions, results in an increase in the matrix stiffness.

On the other hand, an increased free volume between adjacent matrix polymers and a preferential cleavage of bonds by preferential hydrolysis in wet conditions and the subsequent removal of degradation products by water soaking, leads to an irreversible decrease in the cell wall matrix stiffness. This explanation is also consistent with the course of the EMC ratio (EMC of modified wood in relation to the reference EMC) in dependence on the RH at room temperature, based on DVS measurements in **publication III**. Thermal modification of wood in dry state results in a decrease in the EMC ratio with increasing RH, similar to the cell wall matrix stiffening effect of chemical wood modification with cross-linking agents (Yasuda et al. 1994; Himmel and Mai 2015). Thermal modification of wood in wet state, however, leads to an increase in the EMC ratio with increasing relative humidity, which is interpreted as an enhanced expansion of the cell wall nanopores upon moisture uptake. In line with this interpretation, it has been suggested previously that excessive matrix stiffnening during the thermal modification of wood in high temperature and low water vapor pressure environments may almost fully limit the accessibility of sorption sites by expansion of the nanopores (Willems 2014a; Willems 2015).

Dimensional changes

The dimensional stabilization of wood by thermal modification is often explained by the removal of the hydrophilic hemicelluloses. However, the findings of Repellin and Guyonnet (2005) suggest that this explanation is false and that the dimensional stabilization is, rather, achieved by modification of the lignin carbohydrate complex. **Publication III** verifies their findings and suggests that irreversible changes in the cell wall matrix stiffness by modifications of the lignin carbohydrate complex not only affect sorption properties, but are also linked to the reduction of dimensional changes upon drying and capillary re-wetting of thermally modified wood on a macroscopic scale.

Publication III shows that while the maximum swelling is reduced as a function of the mass loss generated when oven-dry wood is thermally modified, there is an increase in maximum swelling if water-saturated wood is thermally modified and water-leached instead. For both, oven-dry and water-saturated wood, the pure removal of hemicelluloses by thermal modification and subsequent water-leaching should lead to reduction in the dry dimensions of the wood after the process, as reported by González-Peña et al. (2009) and illustrated in **Figure 33**. However, the cell wall matrix stiffening effect caused by the modification of the lignin carbohydrate complex during thermal modification of wood in dry state reduces the swelling of the remaining cell wall material. Therefore, the reduction in wet dimensions of the wood after modification is larger than the reduction in dry dimensions, which results in a decrease in maximum swelling (see **Figure 33** A). By surpressing condensation and/or esterification reactions during the thermal modification of wood at high moisture contents, swelling of the remaining cell wall material is not limited by increased cell wall matrix stiffness. Voids previously occupied by hemicelluloses are then filled by water and consequently the reduction in wet dimensions is smaller than the reduction in dry sample dimensions (see **Figure 33** B). This leads to an increase in the maximum swelling when water-saturated samples are thermally modified. The same conclusion was drawn by Burmester (1975) after treating wood with sulfuric acid at elevated temperatures, which lead to an increase in maximum swelling. Interestingly, Hosseinpourpia (2015) impregnated samples of Scots pine sapwood with an aqueous solution of sulfuric acid and exposed the samples to temperatures between 120 and 180°C after a drying step, which resulted in a reduction in maximum swelling. The author considered cell wall matrix stiffening by lignin furan condensation reaction as a possible reason for such a reduction in maximum swelling, which is in line with the explanations above.

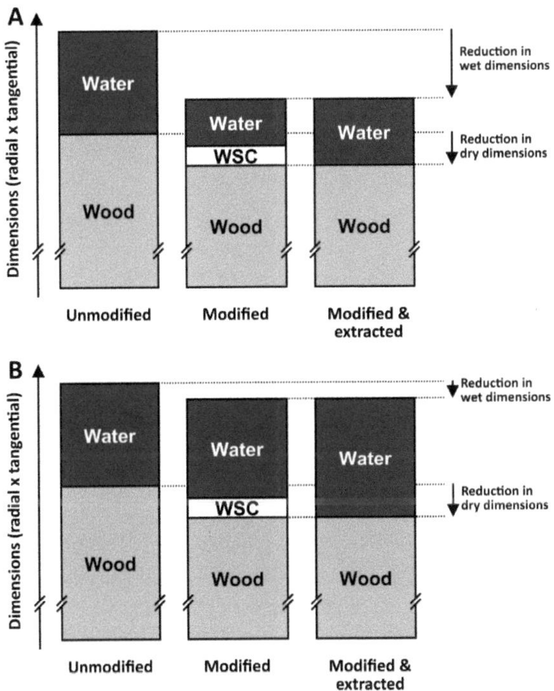

Figure 33: Illustration of changes in wet and dry dimensions by thermal modification and by extraction of water-soluble compounds (WSC). A) Thermal modification of wood in dry state: The reduction in wet dimensions is larger than in dry dimensions. B) Thermal modification of wood in wet state: The reduction in wet dimensions is smaller than in dry dimensions.

There is a theoretical consideration that the impact of water during the process is based on increasing the distance between cell wall polymers, which hinders cross-linking. However, Seborg et al. (1953) used potassium iodide to increase the distance between the polymers instead of water and did not find a significant difference in ASE between oven-dry and potassium iodide-impregnated wood after thermal modification and leaching of the salt. This suggests that the impact of water during the process on the change in maximum swelling is not based on an increase in polymer distance, but rather on a change in reaction rates during thermal modification, as described above.

Although not tested, cell wall matrix stiffening by reversible ultra-structural realignments described in **publication II** are likely to reduce swelling and shrinkage of wood in a similar way as irreversible matrix stiffening by modification of the lignin carbohydrate complex. However, if swelling is generated by conditioning at high RH or by water-soaking, it is expected that the ultra-structural realignments are recovered over time and that the impact on the dimensional stabilization is dimished.

Besides the effect of cell wall matrix stiffening, it was also demonstrated in **publication III** that an additional reduction in maximum swelling is achieved by remaining degradation products within the cell wall that cause a cell wall bulking effect, in line with the findings by Biziks et al. (2015) and Čermák et al. (2015). However, in contrast to the cell wall bulking effect caused by modification agents that are fixed within the wood cell wall after curing, these degradation products are largely extractable with water as a solvent. They can thus be easily removed by several days of water leaching, which diminishes their effect on the maximum swelling during drying and capillary re-wetting.

Mechanical properties

The loss in strength and ductility of wood by thermal modification is strongly linked to the removal of hemicelluloses as the coupling agents between the cell wall polymers and thus strongly correlated to the mass loss. However, changes in the cell wall ultra-structure are likely to contribute to the alteration of the mechanical behaviour (Boonstra et al. 2007a; Hughes et al. 2015). Ultra-structural changes as described in **publication II** as reversible ultra-structural realignments of cell wall matrix polymers, or in **publication III** as irreversible modifications of the lignin-carbohydrate complex, thus may affect wood mechanical properties in addition to their strong impact on sorption and dimensional stability.

The results of a three-point bending test evidence that the reversible, ultra-structural realignments of amorphous matrix polymers described in **publication II** have an effect on the mechanical performance of the modified wood. It is suggested that the ultra-structural realignments hinder the polymer mobility and thus affect the ability of amorphous polymers for plastic flow, during which a rearrangement of the molecular and microscopic structure of materials occurs. This prevents the relaxation of stresses applied by external loads within the inelastic regions of the load-deflection curve. Hence, high stress concentrations are built up and the wood fails at low inelastic deflection. This reduction in deflection is associated with a loss in toughness, indicated by the strong decrease in the inelastic proportion of the work until maximum load in bending. The ultra-structural realignments thus contribute to the loss in toughness and increase in brittleness that is caused by the degradation of hemicelluloses as coupling agents between cell wall polymers (Hughes et al. 2015). This contribution is especially recognizable in view of the fact that thermal modification processes at low process RH that did not lead in any (corrected) mass loss already decreased the inelastic deflection and the inelastic proportion of the work until maximum load to considerable extent. No bending test was performed after exposing the wood to high RH or to water-soaking. It is hypothesized, however, that that the recovery of ultra-structural realignments by such a procedure would result in an increase in inelastic deflection and toughness, similar to its effect on the EMC.

In the framework of this thesis, no mechanical test was performed on wood sampes that were thermally modified in saturated water vapor starting in either oven-dry or water-saturated state, as done in **publication III** for the study of hygroscopicity. Nevertheless, it is hypothesized that reducing the mobility of amorphous cell wall matrix polymers by modification of the lignin carbohydrate complex contributes to the loss in inelastic deflection and toughness in a similar way as described above for reversible ultra-structural realignments. Therefore, thermal modification processes of oven-dry wood should result in a more brittle behavior compared to modification processes of water-saturated wood. A similar effect of wood embrittlement is achieved by the formation of cross-links within the cell wall upon formalization (Burmester 1967) or by the incorporation of a rigid, three-dimensional corset of thermosetting resins such as melamine or phenol formaldehyde into the cell wall (Pittmann et al. 1994; Bicke et al. 2012; Kielmann et al. 2013). It is displayed by a loss in toughness, which is caused by a decrease in deformation in bending or impact bending without changing the maximum load considerably (Bicke et al. 2012; Kielmann et al. 2013).

From this, it follows that while modification and/or realignemts of cell wall matrix polymers in high temperature and low water vapor pressure environments contribute to a brittle behaviour and a loss in toughness of thermally modified wood, the very same ultra-structural changes are (partly) responsible for the improvement in dimensional stability and the reduction in hygroscopicity. The reduced expansion of cell wall nanopores might even play a key role in the improvement in biological durability by inhibiting the penetrability of the wood cell wall for diffusible agents emitted by the fungi in the initial stage of decay (Boonstra et al. 2007c; Zelinka et al. 2014; Hosseinpourpia and Mai 2016; Zelinka et al. 2016). Based on the findings of **publications II and III**, controlling wood drying and the wood moisture content during the process by regulation of the water vapor pressure may thus be a promising approach to optimize the balance between an excessive decrease in the cell wall matrix polymer mobility that results in a strong imbrittlement on the one hand, and a lack of cell wall matrix stiffness that fails to improve the dimensional stability on the other hand. Therefore, the regulation of the water vapor pressure offers additional control over thermal modification processes.

8.2 Surface performance of wood thermally modified in open reactor systems (IV-VI)

The investigations performed in **publications IV-VI** provide explanations why the performance of thermally modified Scots pine and Norway spruce wood in exterior applications does not always meet the expectations derived from the increased resistance against wood decay fungi. These publications study the influence of abiotic factors during surface weathering of thermally modified wood. Since weathering in outdoor conditions is a combination of various factors that function simultaneously (Williams 2005), selected factors, namely repeated capillary wetting and re-drying as well as UV-irradiation in combination with water spray, were simulated under controlled conditions. It was also evaluated if the thermal modification process affects the performance (i.e. the adhesion) of commercially available coating systems.

Wood defects

The development of surface cracks in wood during weathering is a major drawback that decreases the service life of wood. The repeated wetting and re-drying of wood in outdoor conditions caused by changes in the climatic conditions (temperature, RH), rain and sunlight, result in tensile stresses in the outer layers. In thermally modified wood, the water uptake is severely reduced compared to unmodified wood (Metsä-Kortelainen et al. 2006), thus the absolute stress level caused by wetting and re-drying should be lower. For many wood species, however, surface cracking during weathering was not found to be improved or even worsened after thermal modification (Feist and Sell 1987; Jämsä et al. 2000; Flade 2014). Increased photodegradation of thermally modified wood compared to unmodified wood as a cause for the high crack susceptibility during weathering is ruled out by the findings of **publication V**. Furthermore, simulating repeated wetting and re-drying by water submersion and subsequent drying at 50°C in the absence of UV light in **publication IV** resulted in the same increased crack susceptibility of thermally modified Norway spruce and Scots pine as found during weathering.

Publication IV follows the hypothesis of pre-existing cracks formed during the thermal modification process that act as starting points for severe cracks during repeated wetting and re-drying. In line with this hypothesis, an increase in visible surface cracks was indeed found after the thermal modification process and microscopic analysis revealed the formation of microscopic defects. In particular, damages to parenchyma and epithelial cells as well as longitudinal cracks within the earlywood tracheids were detected. A prevention of wood defects formed during the industrial modification process by alterations of the process schedule, i.e. by decreasing the heating rates or by omitting the final conditioning step that involves a water spray, could not be achieved in **publication IV**. This lead to the assumption that the high-temperature drying of wood at the beginning of the process (Mayes and Oksanen 2003) is a major cause for the formation of wood defects during the thermal modification process in an open reactor system. During this initial drying step, the wood will reach the oven-dry state, at which shrinkage stresses caused by anisotropy of the wood reach a maximum (Cheng et al. 2007). It has been claimed previously that avoiding the oven-dry state of the wood by applying an elevated WVP during the process results in less wood defects (Willems 2009; Willems 2010), but no scientific evidence exists. Investigations by Cheng et al. (2007), however, showed that applying elevated RHs in the range between 20 and 80 % within 160 to 180°C does not affect the maximum tangential shrinkage stresses generated upon wood drying. Remarkably, even the use of saturated water vapor (100 % RH) results in significant shrinkage stresses. This indicates that the degradation and modification of the wood constituents create additional stresses within the wood that cannot be avoided by elevated water vapor pressure. However, the cited authors also found that shrinkage stresses during wood drying and associated wood defects are reduced by a short pre-treatment in high-temperature saturated water vapor. This might also be a promising approach to further optimize thermal modification processes with regard to minimizing wood defects.

In addition to factors that are directly related to the thermal modification process, it should be noted that there is a considerable impact of the conventional kiln-drying prior to the modification process as well as of raw material based factors on the formation of cracks in thermally modified wood (**publication IV**). Cracks already present at the wood surface after conventional kiln-drying became more frequent and severe after the thermal modification process. Furthermore, a more horizontal annual ring orientation (tangential) at the board surface resulted in considerably stronger cracking during wetting and re-drying than vertically oriented annual rings (radial), as known for unmodified wood (Flæte et al. 2000). In case of thermal modification processes in open reactor systems, surface cracking of the modified wood in outdoor conditions is thus reduced by a proper selection of the raw material, rather than by further process optimization.

Although the hypothesis of pre-existing cracks in thermally modified wood as a cause for the increased crack susceptibility in outdoor conditions is consistent with the findings of **publication IV**, it lacks verification. This is because **publication IV** is based on comparing qualitative information on microscopic defects observed after the modification process with (semi-) quantitative data on the development of macroscopic surface cracks during repeated wetting and re-drying. The elevated crack susceptibility of thermally modified wood might thus be influenced by additional mechanisms. An additional mechanism could be a limited distribution of stresses induced by repeated wetting and re-drying in thermally modified wood. The preferential degradation of hemicelluloses as the coupling agent between the cellulose microfibrils or between the cellulose and the lignin has been suggested to hinder the distribution of stresses applied by external loads across the cell wall, thus wood fails to function as a composite material (Sweet and Winandy 1999; Winandy and Lebow 2001). Accordingly, the degradation of hemicelluloses has been suggested previously as a major cause for the loss in toughness and ductility caused by thermal modification (Hughes et al. 2015). The reduction in the load sharing capabilities might be further facilitated by (ultra-) structural changes during thermal modification processes, as suggested in **8.1.2**. During the thermal modification in an open reactor system according to the ThermoWood® standards, wood is modified in an oven-dry state, because a high-temperature drying step is applied at the beginning of the process (Mayes and Oksanen 2003). According to the findings in **publication III**, such dry heat conditions result in considerable decrease in the polymer mobility by modification of lignin carbohydrate complex, which might be further enhanced by ultra-structural rearrangements of amorphous matrix polymers upon drying at high temperatures described in **publication II**. In the previous chapter (see **8.1.2**), a low polymer mobility by thermal modification was suggested to not only reduce wood sorption by increasing the cell wall matrix stiffness, but to also hinder the relaxation of stresses applied by external loads. This results in zones of localized stresses in thermally modified wood (González-Peña et al. 2005). In the same way, the reduced load sharing capabilities of thermally modified wood should also affect stresses induced by repeated wetting and re-drying, thus leading in strong cracking even at a reduced absolute level of stresses. If that is the case, then preventing excessive cell wall matrix stiffness by the application of elevated water vapor pressure during the thermal modification might reduce the susceptibility of the wood towards cracking in outdoor conditions compared to processes in open reactor systems.

Photodegradation

Besides the formation of wood defects, photodegradation at the surface is a major drawback during weathering of wood in exterior applications. In **publication V**, the photodegradation of thermally modified wood was assessed on micro-veneers that were subjected to artificial weathering by UV-irradiation and water spray. FT-IR spectra of the micro-veneers collected after different durations of artificial weathering did not indicate differences in the way and the rate at which lignin is degraded by photodegradation in unmodified and thermally modified Norway spruce and Scots pine wood. This finding is in contrast to previous investigations suggesting that the increase in the lignin percentage, the presence of low molecular weight phenols as well as the many alterations of the lignin network during the thermal modification process result in an increased resistance of the modified wood against photodegradation (Ayadi et al. 2003; Huang et al. 2012). The different outcomes might be related to the differences in sample dimensions used for the study of photodegradation. Previous studies (Ayadi et al. 2003; Huang et al. 2012) used solid wood samples, in which weathering products of lignin are largely unleachable in thermally modified wood (Nuopponen et al. 2004). The very small dimensions of the micro-veneers used in **publication V**, however, potentially facilitated the leaching of such weathering products. Low molecular weight phenolic compounds have antioxidant properties (Ahajji et al. 2009) and might thus scavenge light-induced free radicals in wood. Indeed, it has been shown that weathering unmodified wood with UV-light alone is less efficient than a combination of UV-light and water, because of an insufficient removal of weathering products (Anderson et al. 1991). Furthermore, Shen et al (2016) demonstrated recently that UV degradation of thermally modified Scots pine increases if the samples are extracted with an ethanol/toluene misture prior to artificial weathering. Accordingly, the decelerated photodegradation of thermally modified solid wood against photodegradation might thus be assigned to the prevention of leaching of degradation products derived from the lignin by thermal degradation and weathering, rather than to an enhanced resistance against photodegradation of the lignin itself.

In contrast to the FT-IR spectra, changes in the micro-tensile strength measured in conditioned state (20 °C/65 % RH) were not assigned to the degradation of the lignin. Lignin contributes to the tensile strength of wood by limiting the access of water to the cell wall, which helps the wood to retain its strength and stiffness during exposure to moisture (Lagergren et al. 1957). Accordingly, even delignification of micro-veneers does not influence f- or z-strength (finite- or zero-span micro-tensile strength) when measured in conditioned (20 °C/65 % RH) state (Klüppel and Mai 2012). The reductions in f- and z-strength by thermal modification and/or by photodegradation measured in **publication V** are thus assigned to the degradation of carbohydrates. The tensile strength of native wood is strongly determined by the cellulose content and its chain length (Ifju 1964). However, within the temperature range usually applied during thermal modification, degradation of cellulose is limited to the amorphous regions (Tjeerdsma et al. 1998; Boonstra and Tjeerdsma 2006). Furthermore, the degradation of cellulose by thermal modification is outweighed by the degradation of hemicelluloses (Zaman et al. 2000; Alén et al. 2002). In a similar way, photodegradation of carbohydrates in wood are mainly limited to amorphous carbohydrates, i.e. hemicelluloses, since crystalline cellulose is s hardly susceptible to water, chemicals or UV-irradiation (Hon and Feist 1981; Hon and Chang 1984). This predominant degradation of hemicelluloses has been suggested as a major reason for the loss in strength, since it interferes with the load-sharing capabilities of the cell wall matrix that interconnects the cellulose microfibrils (Sweet and Winandy 1999; Winandy and Lebow 2001). In line with this explanation, the loss in f-strength, which is predominantly determined by the cell wall matrix material, was significantly higher after thermal modification or artificial weathering than the loss in z-strength, which is mainly determined by cellulose (Derbyshire et al. 1995; Xie et al. 2007; Klüppel and Mai 2012).

It was further shown in **publication V** that although thermal modification of spruce and pine wood resulted in a loss in f- and z-strength, the strength loss as a function of the duration of the artificial weathering was less facile compared to the unmodified reference. This effect was more pronounced for an increased modification temperature and after 144 h of artificial weathering initial differences in strength loss diminished. It was concluded that the thermal degradation of wood polymers during modification processes decelerate their further degradation during artificial weathering. However, this outcome does not clarify whether the reduced strength loss of by photodegradation is caused by the preferential removal of hemicelluloses, which leads to less carbohydrates being present in thermally modified wood that can be degraded by UV-irradiation, or by the modification of the lignin carbohydrate complex under dry heat conditions that increase the resistance of the remaining carbohydrates against photodegradation. Since removal and chemical alteration of wood components occur simultaneously during thermal modification of wood in oven-dry state (see **publication III**), these effects cannot be differentiated on the basis of **publication V**.

Coating performance

The effect of UV-irradiation as well as repeated wetting and re-drying of the wood surface can be greatly reduced by the application of a surface coating. **Publication IV and V** evidence that such a protection against weathering phenomena can also be necessary in case of thermally modified wood depending on the respective application and the requirements of the end-user. The reduced dimensional stability of thermally modified wood as a substrate should prolong the service life of coatings in exterior applications, because less stresses are generated by movements of the substrate (Podgorski and Roux 1999). Nevertheless, an effective adhesion of the coating to the thermally modified wood surface is a prerequisite for a good performance in exterior applications.

The adhesion strength of two water-borne coating system systems on thermally modified wood, assessed by the pull-off and cross-cut test in **publication VI**, differed depending on the system that was applied. A water-borne acrylate paint performed well with no considerable loss in adhesion strength compared to unmodified wood as substrate. This coincides with Jämsä et al. (2000) and Grüll et al. (2010) who stated that thermally modified wood as substrate does not differ from unmodified wood. However, a water-borne alkyd reinforced acrylate paint resulted in a considerable loss in pull-off strength and an almost complete flaking of the coating during the cross cut test when applied to thermally modified wood.

Microscopic observations of the coating penetration did not indicate a reduction in the mechanical interlocking between thermally modified and unmodified wood as substrate. An increase in the surface hydrophobicity measured by the contact angle of water was not consistent with the loss in adhesion strength. While a significant increase in contact angle of water was only achieved for a severe treatment above 200 °C, a strong loss in adhesion strength was already evident for pine and spruce modified below 200 °C. The results thus confirm previous investigations showing that the wettability of wood by water does not explain the wood-coating interaction sufficiently (Gindl et al. 2004a; Petrič et al. 2007). The increased hydrophobicity of thermally modified wood should thus not be considered as general drawback for the application of water-borne coating systems.

Following the explanations of de Moura et al. (2013) the loss in adhesion strength might be attributed to the strength loss of the substrate by thermal modification. The substrate/coating system would then simply fail at its weakest link, which is indicated by an increase in cohesive failure of wood when thermally modified wood is used as substrate. However, this explanation is inconsistent with the almost unchanged pull-off strength of the water-based acrylate paint, even for wood modified at temperatures above 200 °C. Instead, the adhesion strength must be depending on the specific wood/coating system. Bull and Berasetegui (2006) point out that if the interface bonding is good, a brittle coating on a brittle substrate results in wood/coaiting system that behaves like a brittle bulk material. As reported by Phuong (2007), and shown in **publication II** thermal wood modification results in a significant loss in toughness and in an embrittlement even for low temperatures and mass loss levels. Compared to chain-like acrylate resins, alkyd resins are more brittle due to their three-dimensional structure. The alkyd reinforcement might thus lead to a higher coating brittleness, which explains the low adhesion strength of the water borne alkyd reinforced acrylate paint on thermally modified wood. Using a more ductile coating system without alkyd reinforcement and/or a more ductile wood substrate resulted in higher adhesion strengths.

For some wood species, reduced adhesion strength, increased brittleness of the coating film and peeling of the coating from the wood were attributed to the presence of extractives, i.e. acidic extractives. Extractives can interfere with and hardening reactions of adhesive and coating systems (Anderson 1950; Roffael and Rauch 1974; Wellons et al. 1977; Hse and Kuo 1988). Therefore, the increased acidity of thermally modified spruce and pine observed in **publication VI** might have contributed to the increased brittleness of the alkyd reinforeced acrylate paint. Examples for a dependence of coating and adhesive systems on the changes in acidity of wood by thermal modification already exist in the literature. By preferential removal of carboxylic acids during the thermal modification of European ash (*Fraxinus excelsior* L.), Herrera et al. (2015) found a decreased acidity and an increased adhesion strength of a water-borne coating system that contained an alkyd binder. For PLATO®-treated wood, the low pH of the modified wood surface was suggested as a reason for an insufficient hardening and low adhesion of a phenol-resorcinol-formaldehyde adhesive (Sernek et al. 2008). The strong increase in acidity for thermal modification processes of wood in low temperature, high WVP regimes observed in **publication I** are thus considered as a drawback for an optimal performance of many coating and adhesive systems.

Besides their impact on curing reactions, extractives can also cause the discoloration of coating systems, i.e. by staining from extractives in the knot area (Nussbaum 2004). During thermal modification, native extractives are largely removed from the wood (Nuopponen et al. 2003), as evident from the amounts of ethanol-cyclohexane-soluble extractives, i.e. from the knot area, shown in **publication VI**. However, an increase in the amount of water-soluble extractives and in the acidity of water extracts indicated the formation of new extractives by the thermal degradation of cell wall constituents. Typical thermal degradation products of wood, such as phenolic compounds and aldehydes (Karlsson et al. 2012; Hofmann et al. 2013), contain conjugated double bonds and thus carry the risk of causing discolorations of coating systems. Such a risk is especially high in wood thermally modified in closed reactor systems at elevated pressure, which results in a strong accumulation of degradation products within the wood, as shown in **publication I and III**.

In addition to the adhesion strength, the penetration of coatings into the wood substrate was observed microscopically in **publication VI**. In general, the observed penetration of the two water-borne coatings and the solvent-borne oil followed the same principles reported by Nussbaum (1994) and de Meijer et al. (1998). The penetration of the film-forming water-borne coating systems was limited to the flow into the open tracheid ends, without any visible differences between unmodified and thermally modified wood as substrate. In contrast, a penetration from cell to cell via the pits or the ray cells was evident in case of the non-film forming (penetrating) solvent-borne oil. Regions with an excessively deep penetration of more than 1000 µm were found occasionally in case of thermally modified pine and spruce, which was not the case for unmodified wood. Wood defects caused by the thermal modification process create space for further penetration of solvent-borne coatings into the wood bulk. In this context, the frequent destruction of ray cells found in **publication IV** might especially facilitate the flow of the solvent-borne oil through the rays into deeper regions of thermally modified pine and spruce. An enhanced penetration of a polyurethane adhesive into thermally modified Scots pine was reported recently by Bastani et al. (2016) and the flow through the rays into adjacent tracheids was visualized by X-ray micro-computed tomography. Such an excessive penetration into the wood bulk might necessitate higher oil coverage per area on thermally modified compared to unmodified pine and spruce to ensure a sufficient amount of oil at the wood surface where it is required.

References

Ahajji A, Diouf PN, Aloui F, Elbakali I, Perrin D, Merlin A, George B (2009) Influence of heat treatment on antioxidant properties and colour stability of beech and spruce wood and their extractives. Wood Sci Technol, 43:69–83.

Alén R, Kotilainen R, Zaman A (2002) Thermochemical behavior of Norway spruce (*Picea abies*) at 180-225 °C. Wood Sci Technol, 36:163–171.

Allegretti O, Brunetti M, Cuccui I, Ferrari S, Nocetti M, Terziev N (2012) Thermo-vacuum modification of spruce (*Picea abies*) and fir (*Abies alba* Mill.) wood. BioResources 7:3656–3669.

Allen KW (1987) A review of contemporary views of theories of adhesion. J Adhes, 21:261–277.

Altgen M, Adamopoulos S, Militz H (2015) Wood defects during industrial-scale production of thermally-modified Norway spruce and Scots pine. Wood Mater Sci Eng, doi: 10.1080/17480272.2014.988750

Altgen M, Ala-Viikari J, Hukka A, Tetri T, Militz H (2014) Impact of elevated steam pressure during the thermal modification of Scots pine and Norway spruce. In: Book of abstracts COST Action FP0904 workshop, Skellefteå, Sweden.

Altgen M, Militz H (2015) Effect of temperature and steam pressure during the thermal modification process. In: Hughes M, Rautkari L, Uimonen T, Militz H, Junge B (eds.) Proceedings of the 8th European Conference on Wood Modification, pp. 226-233, Helsinki, Finland.

Altgen M, Willems W, Militz H (2016) Wood degradation affected by process conditions during thermal modification of European beech in a high-pressure reactor system. Eur J Wood Wood Prod, 74:653-662.

Anderson AB (1950) Chemistry of Western pines. Ind Eng Chem, 42:565–569.

Anderson EL, Pawlak Z, Owen NL, Feist WC (1991) Infrared studies of wood weathering. Part I: softwoods. Appl Spectrosc, 45:641–647.

Andersons B, Andersone I, Biziks V, Irbe I, Chirkova J, Sansonetti E, Grinins J, Militz H (2010) Hydrothermal modification for upgrading the durability properties of soft deciduous wood. The International Research Group on Wood Protection, Doc. IRG/WP 10-40494.

Andersson S, Serimaa R, Väänänen T, Paakkari T, Jämsä S, Viitaniemi P (2005) X-ray scattering studies of thermally modified Scots pine (*Pinus sylvestris* L.). Holzforschung, 59:422–427.

Arnold M (2010) Effect of moisture on the bending properties of thermally modified beech and spruce. J Mater Sci 45:669–680.

Askeland DR, Fulay PP, Wright WJ (2010) The science and engineering of materials. Sixth edition. Cengage Learning, Stamford, USA

Awoyemi L, Jones IP (2011) Anatomical explanations for the changes in properties of western red cedar (*Thuja plicata*) wood during heat treatment. Wood Sci Technol, 45:261–267.

Ayadi N, Lejeune F, Charrier F, Charrier B, Merlin A (2003) Color stability of heat-treated wood during artificial weathering. Holz Roh Werkst, 61:221–226.

Barkas WW (1937) Wood-water relationships, part III. Molecular sorption of water by Sitka spruce wood. Proceedings of the Physical Society 49:237-242.

Bastani A, Adamopoulos S, Koddenberg T, Militz H (2016) Study of adhesive bondlines in modified wood with fluorescence microscopy and X-ray micro-computed tomography. Int J Adhes Adhes, 68:351-358.

Bernabei M, Salvatici MC (2016) In situ ESEM observations of spruce wood (*Picea abies* Karst.) during heat treatment. Wood Sci Technol, 50:715–726.

Bicke S, Mai C, Militz H (2012) Modification of beech veneers with low molecular weight phenol formaldehyde for the production of plywood: durability and mechanical properties. In: Jones D, Militz H, Petrič M, Pohleven F, Humar M, Pavlič M (eds.) Proceedings of the 6th European Conference on Wood Modification, pp. 363-366, Ljubljana, Slovenia.

Biziks V, Andersons B, Beļkova Ļ, Kapača E, Militz H (2013) Changes in the microstructure of birch wood after hydrothermal treatment. Wood Sci Technol, 47:717–735.

Biziks V, Andersons B, Sansonetti E, Andersone I, Militz H, Grinins J (2015) One-stage thermo-hydro treatment (THT) of hardwoods: an analysis of form stability after five soaking-drying cycles. Holzforschung, 69:563-571.

Boonstra M, Rijsdijk JF, Sander C, Kegel E, Tjeerdsma B, Militz H, van Acker J, Stevens M (2006a) Microstructural and physical aspects of heat treated wood. Part 1: Softwoods. Maderas Cienc tecnol, 8:193–208.

Boonstra M, Rijsdijk JF, Sander C, Kegel E, Tjeerdsma B, Militz H, van Acker J, Stevens M (2006b) Microstructural and physical aspects of heat treated wood. Part 2: Hardwoods. Maderas Cienc tecnol, 8:209–217.

Boonstra M, Van Acker J, Tjeerdsma B, Kegel E (2007a) Strength properties of thermally modified softwoods and its relation to polymeric structural wood constituents. Ann For Sci, 64:679–690.

Boonstra MJ (2008) A two-stage thermal modification of wood. PhD thesis, University of Ghent, Belgium.

Boonstra MJ, Tjeerdsma B (2006) Chemical analysis of heat treated softwoods. Holz Roh Werkst, 64:204–211.

Boonstra MJ, Van Acker J, Kegel E (2007b) Effect of a two-stage heat treatment process on the mechanical properties of full construction timber. Wood Mater Sci Eng, 2:138–146.

Boonstra MJ, Van Acker J, Kegel E, Stevens M (2007c) Optimisation of a two-stage heat treatment process: durability aspects. Wood Sci Technol 41:31–57.

Borrega M, Kärenlampi PP (2008a) Effect of relative humidity on thermal degradation of Norway spruce (*Picea abies*) wood. J Wood Sci, 54:323–328.

Borrega M, Kärenlampi PP (2008b) Mechanical behavior of heat-treated spruce (*Picea abies*) wood at constant moisture content and ambient humidity. Holz Roh Werkst, 66:63–69.

Borrega M, Kärenlampi PP (2010) Hygroscopicity of heat-treated Norway spruce (*Picea abies*) wood. Eur J Wood Wood Prod, 68:233–235.

Bourgois J, Guyonnet R (1988) Characterization and analysis of torrefied wood. Wood Sci Technol, 22:143–155.

Boyd JD (1982) An anatomical explanation for visco-elastic and mechano-sorptive creep in wood, and effects of loading rate on strength. In: Baas P (ed) New perspectives in wood anatomy. Martinus Nijhoff publishers, La Hague

Bucur V (2011) Delamination in wood, wood products and wood-based composites. Springer, New York

Bull SJ, Berasetegui EG (2006) An overview of the potential of quantitative coating adhesion measurement by scratch testing. In: Sinha, SK (ed.) Scratching of materials and applications. Tribology and interface engineering series, No. 51, Elsevier B.V., Amsterdam.

Burmester A (1967) Versuche zur Behandlung von Holz mit monomerem Formaldehyd-Gas unter Verwendung von Gamma-Strahlen. Holzforschung, 21:13-20.

Burmester A (1973) Einfluß einer Wärme-Druck-Behandlung halbtrockenen Holzes auf seine Formbeständigkeit. Holz Roh Werkst, 31:237–243.

Burmester A (1975) Zur Dimensionsstabilisierung von Holz. Holz Roh Werkst, 33:333–335.

Candelier K, Dumarçay S, Pétrissans A, Desharnais L, Gérardin P, Pétrissans M (2013) Comparison of chemical composition and decay durability of heat treated wood cured under different inert atmospheres: nitrogen or vacuum. Polym Degrad Stabil, 98:677–681.

Carrasco F, Roy C (1992) Kinetic study of dilute-acid prehydrolysis of xylan-containing biomass.Wood Sci Technol, 26:189–208.

CEN/TS 15083-1: (2005) Durability of wood and wood-based products. Determination of the natural durability of solid wood against food-destroying fungi, test methods. Basidiomycetes. CEN - European Comitee for Standardization, Brussels, Belgium

Čermák P, Rautkari L, Horáček P, Saake B, Rademacher P, Sablík P (2015) Analysis of dimensional stability of thermally modified wood affected by re-wetting cycles. BioResources, 10:3242–3253.

Chang S, Hon D, Feist W (1982) Photodegradation and photoprotection of wood surfaces. Wood Fiber Sci, 14:104–117.

Chaouch M, Dumarçay S, Pétrissans A, Pétrissans M, Gérardin P (2013) Effect of heat treatment intensity on some conferred properties of different European softwood and hardwood species. Wood Sci Technol, 47:663–673.

Cheng W, Morooka T, Wu Q, Liu Y (2007) Characterization of tangential shrinkage stresses of wood during drying under superheated steam above 100°C. Forest Prod J, 57:39–43.

Chow SZ, Pickles KJ (1971) Thermal softening and degradation of wood and bark. Wood Fiber Sci, 3:166–178.

Crawshaw J, Cameron RE (2000) A small angle X-ray scattering study of pore structure in Tencel® cellulose fibres and the effects of physical treatments. Polymer, 41:4691–4698.

Daniel G, Nilsson T, Pettersson B (1989) Intra- and extracellular localization of lignin peroxidase during the degradation of solid wood and wood fragments by *Phanerochaete chrysosporium* by using transmission electron microscopy and immuno-gold labeling. Appl Environ Microbiol, 55:871–881.

de Meijer M, Thurich K, Militz H (1998) Comparative study on penetration characteristics of modern wood coatings. Wood Sci Technol, 32:347–365.

de Moura LF, Brito JO, Nolasco AM, Uliana LR, Bolzon De Muniz GI (2013) Evaluation of coating performance and color stability on thermally rectified *Eucalyptus grandis* and *Pinus caribaea* var. *hondurensis* wood. Wood Res-Slovakia, 58:231–242.

Demirbaş A (2000) Mechanisms of liquefaction and pyrolysis reactions of biomass. Energ Convers Manage, 41:633–646.

Derbyshire H, Miller E, Turkulin H (1995) Investigations into the photodegradation of wood using microtensile testing – Part 1: The application of microtensile testing to measurement of photodegradation rates. Holz Roh Werkst, 53:339–345.

Derbyshire H, Miller ER (1981) The photodegradation of wood during solar irradiation – Part 1. Effects on the structural integrity of thin wood strips. Holz Roh Werkst, 39:341–350.

Derbyshire H, Miller ER, Turkulin H (1996) Investigations into the photodegradation of wood using microtensile testing – Part 2: An investigation of the changes in tensile strength of different softwood species during natural weathering. Holz Roh Werkst, 54:1–6.

Derbyshire H, Miller ER, Turkulin H (1997) Investigations into the photodegradation of wood using microtensile testing – Part 3: The influence of temperature on photodegradation rates. Holz Roh Werkst, 55:287–291.

DIN 52186: (1978) Prüfung von Holz - Biegeversuch. DIN - Deutsches Institut für Normung e.V.

Ding T, Gu L, Li T (2011a) Influence of steam pressure on physical and mechanical properties of heat-treated Mongolian pine lumber. Eur J Wood Wood Prod, 69:121–126.

Ding T, Gu LB, Liu X (2011b) Influence of steam pressure on chemical changes of heat-treated Mongolian pine wood. BioResources, 6:1880–1889.

Dinwoodie JM (1989) Wood: nature's cellular, polymeric fibre-composite. Institute of Metals, London

Du X, Gellerstedt G, Li J (2013) Universal fractionation of lignin–carbohydrate complexes (LCCs) from lignocellulosic biomass: an example using spruce wood. Plant J, 74:328–338.

Dubey MK, Pang S, Walker J (2012) Changes in chemistry, color, dimensional stability and fungal resistance of *Pinus radiata* D. Don wood with oil heat-treatment. Holzforschung, 66:49–57.

DuBois M, Gilles KA, Hamilton JK, Rebers PA, Smith F (1956) Colorimetric method for determination of sugars and related substances. Anal Chem, 28:350–356.

Duchesne I, Hult E, Molin U, Daniel G, Iversen T, Lennholm H (2001) The influence of hemicellulose on fibril aggregation of kraft pulp fibres as revealed by FE-SEM and CP/MAS ^{13}C-NMR. Cellulose, 8:103–111.

Eaton RA, Hale MDC (1993) Wood: Decay, pests and protection. Chapmann & Hall, London, UK

Eichhorn SJ, Young RJ (2001) The Young's modulus of a microcrystalline cellulose. Cellulose, 8:197–207.

Emsley AM, Stevens GC (1994) Kinetics and mechanisms of the low-temperature degradation of cellulose. Cellulose, 1:26–56.

EN 113: (1996) Wood preservatives - Test method for determining the protective effectiveness against wood destroying basidiomycetes - Determination of the toxic values. CEN - European Comitee for Standardization, Brussels, Belgium

EN 252: (2014) Field test method for determining the relative protective effectiveness of a wood preservative in ground contact. CEN - European Comitee for Standardization, Brussels, Belgium

EN 335-1: (2006) Durabiltity of wood and wood-based products - Definitions of use classes. General. CEN - European Comitee for Standardization, Brussels, Belgium

EN 350-1: (1994) Durability of wood and wood-based products. Natural durability of solid wood. Guide to the principles of testing and classification of natural durability of wood. CEN - European Comitee for Standardization, Brussels, Belgium

EN 927-6: (2008) Paints and varnishes - Coating materials and coating systems for exterior wood. Exposure of wood coatings to artificial weathering using fluorescent UV lamps and water. CEN - European Comitee for Standardization, Brussels, Belgium

Endo K, Obataya E, Zeniya N, Matsuo M (2016) Effects of heating humidity on the physical properties of hydrothermally treated spruce wood. Wood Sci Technol, doi: 10.1007/s00226-016-0822-4

Engelhardt F (1979) Untersuchungen über die Wasserdampfsorption durch Buchenholz im Temperaturbereich von 110 bis 170°C. Holz Roh Werkst, 37:99–112.

Engelund ET, Thygesen Lisbeth G, Hoffmeyer P (2010) Water sorption in wood and modified wood at high values of relative humidity. Part 2: Appendix. Theoretical assessment of the amount of capillary water in wood microvoids. Holzforschung, 64:325-330.

Engelund ET, Thygesen LG, Svensson S, Hill CAS (2013) A critical discussion of the physics of wood-water interactions. Wood Sci Technol 47:141–161.

Esteves B, Graça J, Pereira H (2008) Extractive composition and summative chemical analysis of thermally treated eucalypt wood. Holzforschung, 62:344–351.

Esteves B, Marques AV, Domingos I, Pereira H (2007) Influence of steam heating on the properties of pine (*Pinus pinaster*) and eucalypt (*Eucalyptus globulus*) wood. Wood Sci Technol, 41:193–207.

Esteves B, Videira R, Pereira H (2011) Chemistry and ecotoxicity of heat-treated pine wood extractives. Wood Sci Technol, 45:661–676.

Esteves BM, Pereira HM (2009) Wood modification by heat treatment: A review. BioResources, 4:370–404.

Evans AG, Drory MD, Hu MS (1988) The cracking and decohesion of thin films. J Mater Res, 3: 1043-1049.

Fahlén J, Salmén L (2003) Cross-sectional structure of the secondary wall of wood fibers as affected by processing. J Mater Sci, 38:119–126.

Faix O (1991) Classification of lignins from different botanical origins by FT-IR spectroscopy. Holzforschung, 45:21-28.

FAO (2012) State of the world's forests 2012. Food and Agriculture Organization of the United Nations, Rome

Feist WC, Sell J (1987) Weathering behavior of dimensionally stabilized wood treated by heating under pressure of nitrogen gas. Wood Fiber Sci, 19:183–195.

Fengel D (1966) On changes of wood and its components within temperature range up to 200°C – Part I. Hot- and cold-water extracts of thermally treated spruce wood. Holz Roh Werkst, 24:9-14.

Fengel D (1967) On the changes of the wood and its components within the temperature range up to 200°C – Part IV: The behaviour of cellulose in spruce wood under thermal treatment. Holz Roh Werkst, 25:102–111.

Fengel D, Wegener G (1984) Wood: Chemistry, ultrastructure, reactions. Walter de Gruyter, Berlin

Flade P (2014) Rissbildung an TMT: Problem und Lösungsansätze. In: Proceedings of the 8th European TMT-Workshop, pp. 22–30, Dresden, Germany.

Flæte PO, Høibø OA, Fjærtoft F, Nilsen TN (2000) Crack formation in unfinished siding of aspen (*Populus tremula* L.) and Norway spruce (*Picea abies* (L.) Karst.) during accelerated weathering. Holz Roh Werkst, 58:135–139.

Forest Products Laboratory (1957) Shrinking and swelling of wood in use. United States Department of Agriculture, Forest Service, Forest Products Laboratory, Madison, WI

Fromm J (2013) Xylem development in trees: from cambial divisions to mature wood cells. In: Fromm J (ed) Cellular aspects of wood formation. Springer, Berlin Heidelberg

Gardner DJ, Generalla NC, Gunnells DW, Wolcott MP (1991) Dynamic wettability of wood. Langmuir 7:2498–2502.

Garrote G, Domínguez H, Parajó JC (2001) Study on the deacetylation of hemicelluloses during the hydrothermal processing of Eucalyptus wood. Holz Roh Werkst, 59:53–59.

Garrote G, Domínguez H, Parajó JC (1999) Hydrothermal processing of lignocellulosic materials. Holz Roh Werkst, 57:191–202.

Gérardin P, Petrič M, Petrissans M, Lambert J, Ehrhardt JJ (2007) Evolution of wood surface free energy after heat treatment. Polym Degrad Stabil, 92:653–657.

Giebeler E (1983) Dimensional stabilization of wood by moisture-heat-pressure-treatment. Holz Roh Werkst, 41:87–94.

Gindl M, Reiterer A, Sinn G, Stanzl-Tschegg SE (2004a) Effects of surface ageing on wettability, surface chemistry, and adhesion of wood. Holz Roh Werkst 62:273–280.

Gindl W, Gupta HS, Schöberl T, Lichtenegger HC, Fratzl P (2004b) Mechanical properties of spruce wood cell walls by nanoindentation. Appl Phys A, 79:2069–2073.

González-Peña MM, Breese MC, Hale MDC (2005) Studies on the relaxation of heat-treated wood. In: Militz H, Hill C (eds) Proceedings of the 2nd European Conference on Wood Modification, pp. 87–90, Göttingen, Germany.

González-Peña MM, Curling SF, Hale MDC (2009) On the effect of heat on the chemical composition and dimensions of thermally-modified wood. Polym Degrad Stabil, 94:2184–2193.

González-Peña MM, Hale MDC (2007) The relationship between mechanical performance and chemical changes in thermally modified wood. In: Hill CAS, Jones D, Militz H, Ormondroyd GA (eds.) Proceedings of the 3rd European Conference on Wood Modification, pp. 169–172, Cardiff, UK.

Goodell B, Jellison J, Liu J, Daniel G, Paszsczynski A, Fekete F, Krishnamurthy S, Jun L, Xu G (1997) Low molecular weight chelators and phenolic compounds isolated from wood decay fungi and their role in the fungal biodegradation of wood. J Biotechnol, 53:133–162.

Griffin DM (1977) Water potential and wood-decay fungi. Annu Rev of Phytopathol, 15:319–329.

Grüll G, Podgorski L, Truskaller M, Spitaler I, Georges V, Bollmus S, Steitz A (2010) Performance of selected types of coated and uncoated modified wood in artificial and natural weathering. The International Research Group on Wood Protection, Doc. IRG/WP 10-40510.

Hakkou M, Pétrissans M, El Bakali I, Gérardin P, Zoulalian A (2005) Wettability changes and mass loss during heat treatment of wood. Holzforschung 59:35–37.

Hakkou M, Pétrissans M, Gérardin P, Zoulalian A (2006) Investigations of the reasons for fungal durability of heat-treated beech wood. Polym Degrad Stabil, 91:393–397.

Hammel KE, Kapich AN, Jensen JR KA, Ryan ZC (2002) Reactive oxygen species as agents of wood decay by fungi. Enzyme Microb Tech, 30:445–453.

Hanhijärvi A, Wahl P, Räsänen J, Silvennoinen R (2003) Observation of development of microcracks on wood surface caused by drying stresses. Holzforschung, 57:561–565.

Harrington KJ, Higgins HG, Michell AJ (1964) Infrared spectra of *Eucalyptus regnans* F. Muell. and *Pinus radiata* D. Don. Holzforschung, 18:108-113.

Herrera R, Muszyńska M, Krystofiak T, Labidi J (2015) Comparative evaluation of different thermally modified wood samples finishing with UV-curable and waterborne coatings. Appl Surf Sci, 357:1444–1453.

Hill CAS (2006) Wood modification: Chemical, thermal and other processes. John Wiley & Sons, Ltd., Chichester

Hill CAS, Jones D (1996) The dimensional stabilisation of Corsican pine sapwood by reaction with carboxylic acid anhydrides. The effect of chain length. Holzforschung, 50:457–462.

Hill CAS, Keating BA, Jalaludin Z, Mahrdt E (2012a) A rheological description of the water vapour sorption kinetics behaviour of wood invoking a model using a canonical assembly of Kelvin-Voigt elements and a possible link with sorption hysteresis. Holzforschung, 66:35–47.

Hill CAS, Ramsay J, Keating B, Laine K, Rautkari L, Hughes M, Constant B (2012b) The water vapour sorption properties of thermally modified and densified wood. J Mater Sci, 47:3191–3197.

Hillis WE (1984) High-temperature and chemical effects on wood stability – Part 1: General considerations. Wood Sci Technol, 18:281–293.

Himmel S, Mai C (2015) Effects of acetylation and formalization on the dynamic water vapor sorption behavior of wood. Holzforschung, 69:633-643.

Hingston J, Collins C, Murphy R, Lester J (2001) Leaching of chromated copper arsenate wood preservatives: a review. Environ Pollut, 111:53–66.

Hoffmeyer P, Engelund ET, Thygesen LG (2011) Equilibrium moisture content (EMC) in Norway spruce during the first and second desorptions. Holzforschung 65:875-882.

Hoffmeyer P, Jensen SK, Jones D, et al (2003) Sorption properties of steam treated wood and plant fibers. In: van Acker J, Hill CAS (eds) Proceedings of the 1st European Conference on Wood Modification, pp. 177-189, Ghent, Belgium.

Hofmann T, Rétfalvi T, Albert L, Niemz P (2008) Investigation of the chemical changes in the structure of wood thermally modified within a nitrogen atmosphere autoclave. Wood Res-Slovakia, 53:85–98.

Hofmann T, Wetzig M, Rétfalvi T, Sieverts T, Bergemann H, Niemz P (2013) Heat-treatment with the vacuum-press dewatering method: chemical properties of the manufactured wood and the condensation water. Eur J Wood Wood Prod, 71:121–127.

Hon DNS, Chang ST (1984) Surface degradation of wood by ultraviolet light. J Poly Sci Pol Chem, 22:2227–2241.

Hon DNS, Feist WC (1981) Free-radical formation in wood – The role of water. Wood Sci, 14:41–48.

Hon DNS, Ifju G (1978) Measuring penetration of light into wood by detection of photoinduced free-radicals. Wood Sci, 11:118–127.

Hosseinpourpia R (2015) Influence of impregnation, chemical and thermal wood modification on properties related to sorption, tensile strength and resistance to Fenton's reagent. PhD thesis, Georg-August University of Göttingen, Germany

Hosseinpourpia R, Adamopoulos S, Mai C (2016) Dynamic vapour sorption of wood and holocellulose modified with thermosetting resins. Wood Sci Technol, 50:165-178.

Hosseinpourpia R, Mai C (2016) Mode of action of brown rot decay resistance of thermally modified wood: resistance to Fenton's reagent. Holzforschung, 70:691-697.

Hse CY, Kuo MI (1988) Influence of extractives on wood gluing and finishing – a review. Forest Prod J, 38:52–56.

Huang XA, Kocaefe D, Kocaefe Y, Boluk Y, Pichette A (2012) A spectrocolorimetric and chemical study on color modification of heat-treated wood during artificial weathering. Appl Surf Sci, 258:5360–5369.

Hughes M, Hill C, Pfriem A (2015) The toughness of hygrothermally modified wood – a review. Holzforschung, 69:851-862.

Ibbett R, Gaddipati S, Davies S, Hill S, Tucker G (2011) The mechanisms of hydrothermal deconstruction of lignocellulose: New insights from thermal-analytical and complementary studies. Bioresource Technol, 102:9272–9278.

Ifju G (1964) Tensile strength behavior as a function of cellulose in wood. Forest Prod J, 14:366–372.

Inari GN, Pétrissans M, Lambert J, Ehrhardt JJ, Gérardin P (2006) XPS characterization of wood chemical composition after heat-treatment. Surf Interface Anal, 38:1336–1342.

Inari GN, Pétrissans M, Pétrissans A, Gérardin P (2009) Elemental composition of wood as a potential marker to evaluate heat treatment intensity. Polym Degrad Stabil, 94:365–368.

Ishikawa A, Kuroda N, Kato A (2004) In situ measurement of wood moisture content in high-temperature steam. J Wood Sci, 50:7–14.

ISO 4628-4: (1982) Paints and varnishes - Evaluation of degradation of coatings. Designation of quantity and size of defects, and of intensity of uniform changes in appearance. Assessment of degree of cracking. International Organization for Standardization, Geneva, Switzerland

Jalaludin Z, Hill CAS, Xie Y, Samsi HW, Husain H, Awang K. Curling SF (2010) Analysis of the water vapour sorption isotherms of thermally modified acacia and sesendok. Wood Mater Sci Eng, 5:194–203.

Jämsä S, Ahola P, Viitaniemi P (2000) Long-term natural weathering of coated ThermoWood. Pigm Resin Technol, 29:68–74.

Jermer J, Edlund ML (1990) Will political initiatives stop the use of preservative-treated wood in Sweden? International Research Group on Wood Protection, Doc. IRG/WP/3578.

Johansson D, Sehlstedt-Persson M, Morén T (2006) Effect of heat treatment on capillary water absorption of heat-treated pine, spruce and birch. In: Kurjatko S, Kúdela J, Lagaňa (eds.) Proceedings of the 5th IUFRO Symposium Wood Structure and Wood Properties, pp. 251-255 Sliač, Slovenia.

Kamdem DP, Pizzi A, Jermannaud A (2002) Durability of heat-treated wood. Holz Roh Werkst, 60:1–6.

Kamdem DP, Pizzi A, Triboulot MC (2000) Heat-treated timber: potentially toxic byproducts presence and extent of wood cell wall degradation. Holz Roh Werkst, 58:253–257.

Karlsson O, Torniainen P, Dagbro O, Granlund K, Morén T (2012) Presence of water-soluble compounds in thermally modified wood: carbohydrates and furfurals. BioResources, 7:3679–3689.

Kataoka Y, Kiguchi M (2001) Depth profiling of photo-induced degradation in wood by FT-IR microspectroscopy. J Wood Sci, 47:325–327.

Kesik HI, Akyildiz MH (2015) Effect of heat treatment on the adhesion strength of water based wood varnishes. Wood Res-Slovakia, 60: 987-994.

Kielmann B, Militz H, Mai C (2013) Strength changes in ash, beech and maple wood modified with a N-methylol melamine compound and a metal complex dye. Wood Res-Slovakia, 58:343-350.

Klüppel A, Mai C (2012) Effect of lignin and hemicelluloses on the tensile strength of micro-veneers determined at finite span and zero span. Holzforschung, 66:493–496.

Kol HŞ, Sefil Y (2011) The thermal conductivity of fir and beech wood heat treated at 170, 180, 190, 200, and 212°C. J Appl Polym Sci, 121:2473–2480.

Kollmann F (1962) Eine Gleichung der Sorptionsisotherme. Die Naturwissenschaften, 206–207.

Kollmann F, Fengel D (1965) Änderungen der chemischen Zusammensetzung von Holz durch thermische Behandlung. Holz Roh Werkst, 23:461–468.

Kollmann F, Schmidt E, Kufner M, Fengel D, Scheider A (1969) The alteration of texture and properties of wood by mechanical and thermal stresses. Holz Roh Werkst, 27:407–425.

Kollmann F, Schneider A (1963) Über das Sorptionsverhalten wärmebehandelter Hölzer. Holz Roh Werkst, 21:77–85.

Korkut S, Akgül M, Dündar T (2008a) The effects of heat treatment on some technological properties of Scots pine (*Pinus sylvestris* L.) wood. Bioresource Technol, 99:1861–1868.

Korkut S, Kök MS, Korkut DS, Gürleyen T (2008b) The effects of heat treatment on technological properties in Red-bud maple (*Acer trautvetteri* Medw.) wood. Bioresource Technol, 99:1538–1543.

Koshijima T, Watanabe T (2003) Association between lignin and carbohydrates in wood and other plant tissues. Springer-Verlag, Berlin

Kotilainen R (2000) Chemical changes in wood during heating at 150-160°C. PhD thesis, University of Jyväskylä, Finland

Kotilainen RA, Toivanen TJ, Alén RJ (2000) FTIR monitoring of chemical changes in softwood during heating. J Wood Chem Technol, 20:307–320.

Kubojima Y, Okano T, Ohta M (2000) Bending strength and toughness of heat-treated wood. J Wood Sci, 46:8–15.

Kubojima Y, Suzuki Y, Tonosaki M, Ishikawa A (2003) Moisture content of green wood in high temperature water vapor. Holzforschung, 57:634–638.

Lagergren S, Rydholm S, Stockman L (1957) Studies on the interfibre bonds of wood. Part 1: Tensile strength of wood after heating, swelling and delignification. Sven Papperstidn, 60:632–644.

Lamb FM (1992) Splits and cracks in wood. In: Proceedings of the 43rd Meeting of the Western Dry Kiln Association, pp. 16-24, Reno, Nevada.

Larsson PT, Wickholm K, Iversen T (1997) A CP/MAS[13]C NMR investigation of molecular ordering in celluloses. Carbohyd Res, 302:19–25.

Lenth CA, Kamke FA (2001) Equilibrium moisture content of wood in high-temperature pressurized environments. Wood Fiber Sci, 33:104–118.

LeVan SL, Ross RJ, Winandy JE (1990) Effects of fire retardant chemicals on the bending properties of wood at elevated temperatures. Res Pap FPL-RP498

Li J, Henriksson G, Gellerstedt G (2007) Lignin depolymerization/repolymerization and its critical role for delignification of aspen wood by steam explosion. Bioresource Technol, 98:3061–3068.

Majano-Majano A, Hughes M, Fernandez-Cabo J (2012) The fracture toughness and properties of thermally modified beech and ash at different moisture contents. Wood Sci Technol, 46:5–21.

Majka J, Czajkowski Ł, Olek W (2016) Effects of cyclic changes in relative humidity on the sorption hysteresis of thermally modified spruce wood. BioResources, 11:5265–6275.

Manninen A-M, Pasanen P, Holopainen JK (2002) Comparing the VOC emissions between air-dried and heat-treated Scots pine wood. Atmos Environ, 36:1763–1768.

Mayes D, Oksanen O (2003) ThermoWood Handbook. Finnish ThermoWood Association, Helsinki, Finland

Metsä-Kortelainen S (2011) Differences between sapwood and heartwood of thermally modified Norway spruce (*Picea abies*) and Scots pine (*Pinus sylvestris*) under water and decay exposure. PhD thesis, VTT Technical Research Centre of Finland, Helsinki, Finland.

Metsä-Kortelainen S, Antikainen T, Viitaniemi P (2006) The water absorption of sapwood and heartwood of Scots pine and Norway spruce heat-treated at 170 °C, 190 °C, 210 °C and 230 °C. Holz Roh Werkst, 64:192–197.

Metsä-Kortelainen S, Paajanen L, Viitanen H (2011) Durability of thermally modified Norway spruce and Scots pine in above-ground conditions. Wood Mater Sci Eng, 6:163-169.

Metsä-Kortelainen S, Viitanen H (2009) Decay resistance of sapwood and heartwood of untreated and thermally modified Scots pine and Norway spruce compared with some other wood species. Wood Mater Sci Eng, 4:105–114.

Metsä-Kortelainen S, Viitanen H (2012) Wettability of sapwood and heartwood of thermally modified Norway spruce and Scots pine. Eur J Wood Wood Prod, 70:135–139.

Militz H, Altgen M (2014) Processes and properties of thermally modified wood manufactured in Europe. In: Schultz TP, Goodell B, Nicholas DD (eds.) Deterioration and protection of sustainable biomaterials. ACS Symposium Series 1158, Chap. 16, pp 269–285, Oxford University Press.

Militz H, Beckers EPJ, Homan WJ (1997) Modification of solid wood: research and practical application. The International Research Group on Wood Preservation, Doc. IRG/WP 97-40098.

Miller GL (1959) Use of dinitrosalicylic acid reagent for determination of reducing sugar. Anal Chem, 31:426–428.

Minor JL (1994) Hornification - Its origin and meaning. Prog Pap Recycl, 3:93–95.

Mitchell PH (1988) Irreversible property changes of small Loblolly-pine specimens heated in air, nitrogen, or oxygen. Wood Fiber Sci, 20:320–335.

Mohareb A, Sirmah P, Pétrissans M, Gérardin P (2012) Effect of heat treatment intensity on wood chemical composition and decay durability of *Pinus patula*. Eur J Wood Wood Prod, 70:519–524.

Morén T (1993) Creep, deformation and moisture redistribution during air convective wood drying and conditioning. PhD thesis, Luleå University of Technology, Sweden.

Mundigler N, Rettenbacher M (2005) Natwood technology - A material thermal wood modification. In: Militz H, Hill C (eds) Proceedings of the 2nd European Conference on Wood Modification, pp. 270–275, Göttingen, Germany.

Nguyen T, Johns WE (1979) The effects of aging and extraction on the surface free energy of Douglas fir and redwood. Wood Sci Technol, 13:29–40.

Norrström H (1969) Light absorbing properties of pulp and pulp components. Sven Papperstidn Nord Cellul, 72:25–38.

Nuopponen M (2005) FT-IR and UV Raman spectroscopic studies on thermal modification of Scots pine wood and its extractable compounds. PhD-thesis, Helsinki University of Technology

Nuopponen M, Vuorinen T, Jämsä S, Viitaniemi P (2005) Thermal modifications in softwood studied by FT-IR and UV Resonance Raman spectroscopies. J Wood Chem Technol, 24:13–26.

Nuopponen M, Vuorinen T, Jamsa S, Viitaniemi P (2003) The effects of a heat treatment on the behaviour of extractives in softwood studied by FTIR spectroscopic methods. Wood Sci Technol, 37:109–115.

Nuopponen M, Wikberg H, Vuorinen T, Maunu SL, Jämsä S, Viitaniemi P (2004) Heat-treated softwood exposed to weathering. J Appl Polym Sci, 91:2128–2134.

Nussbaum RM (1999) Natural surface inactivation of Scots pine and Norway spruce evaluated by contact angle measurements. Holz Roh Werkst, 57:419–424.

Nussbaum RM (2004) The effect of processing and treatment parameters on the discoloration of painted joinery due to resin exudation from knots. Surf Coat Int Pt B Coat Trans, 87:181–186.

Nussbaum RM (1994) Penetration of water-borne alkyd emulsions and solvent-borne alkyds into wood – autodiographic and SEM-EDXA studies. Holz Roh Werkst, 52:389–393.

Obataya E (2007) Characteristics of aged wood and Japanese traditional coating technology for wood protection. Actes de la journée d'étude Conserver aujourd'hui: les vieillissements du bois – Cité de la Musique, pp. 26-43.

Obataya E, Higashihara T, Tomita B (2002) Hygroscopicity of heat-treated wood III. Effect of steaming on the hygroscopicity of wood. Mokuzai Gakkaishi, 48:348–355.

Obataya E, Tomita B (2002) Hygroscopicity of heat-treated wood II. Reversible and irreversible reductions in the hygroscopicity of wood due to heating. Mokuzai Gakkaishi, 48:288–295.

Olek W, Majka J, Czajkowski Ł (2013) Sorption isotherms of thermally modified wood. Holzforschung, 67:183-191.

Pandey KK, Theagarajan KS (1997) Analysis of wood surfaces and ground wood by diffuse reflectance (DRIFT) and photoacoustic (PAS) Fourier transform infrared spectroscopic techniques. Holz Roh Werkst, 55:383–390.

Peters J, Pfriem A, Horbens M, Fischer S, Wagenführ A (2009) Emissions from thermally modified beech wood, their reduction by solvent extraction and fungicidal effect of the organic solvent extracts. Wood Mater Sci Eng, 4:61–66.

Petrič M, Knehtl B, Krause A, Militz H, Pavlič M, Pétrissans M, Rapp A, Tomažič M, Welzbacher C, Gérardin P (2007) Wettability of waterborne coatings on chemically and thermally modified pine wood. J Coat Technol Res, 4:203–206.

Pétrissans M, Gérardin P, Bakali I, Serraj M (2003) Wettability of heat-treated wood. Holzforschung, 57:301–307.

Phuong L, Shida S, Saito Y (2007) Effects of heat treatment on brittleness of *Styrax tonkinensis* wood. J Wood Sci, 53:181–186.

Pittman CU, Kim MG, Nicholas DD, Wang L, Kabir FRA, Schultz TP, Ingram LL (1994) Wood enhancement treatments - I. Impregnation of Southern yellow pine with melamine-formaldehyde and melamine-ammeline-formaldehyde resins. J Wood Chem Technol, 14:577-603.

Pizzi A (1992) A brief, non-mathematical review of adhesion theories as regards their applicability to wood. Holzforsch Holzverw, 44:6–11.

Podgorski L, Roux M (1999) Wood modification to improve the durability of coatings. Surf Coat Int, 82:590–596.

Poncsak S, Kocaefe D, Simard F, Pichette A (2009) Evolution of extractive composition during thermal treatment of Jack pine. J Wood Chem Technol, 29:251–264.

Popescu C-M, Hill CAS (2013) The water vapour adsorption–desorption behaviour of naturally aged *Tilia cordata* Mill. wood. Polym Degrad Stabil, 98:1804–1813.

Popper R, Bariska M (1972) Acylation of wood - Part I: The sorption behavior of water vapour. Holz Roh Werkst, 30:289–294.

Popper R, Bariska M (1973) Acylation of wood - Part II: Thermodynamics of water vapour sorption. Holz Roh Werkst, 31:65–70.

Popper R, Bariska M (1975) Acylation of wood - Part III: Swelling and shrinking behaviour. Holz Roh Werkst, 33:415–419.

Popper R, Niemz P, Eberle G (2005) Investigations on the sorption and swelling properties of thermally treated wood. Holz Roh Werkst, 63:135–148.

Raczkowski J (1980) Seasonal effects on the atmospheric corrosion of spruce micro-sections. Holz Roh Werkst, 38:231–234.

Rautkari L, Hill CAS (2014) Effect of initial moisture content on the anti-swelling efficiency of thermally modified Scots pine sapwood treated in a high-pressure reactor under saturated steam. Holzforschung, 68:323–326.

Rautkari L, Hill CS, Curling S, Jalaludin Z, Ormondroyd G (2013) What is the role of the accessibility of wood hydroxyl groups in controlling moisture content? J Mater Sci, 48:6352–6356.

Rautkari L, Honkanen J, Hill CAS, Ridley-Ellis D, Hughes M (2014) Mechanical and physical properties of thermally modified Scots pine wood in high pressure reactor under saturated steam at 120, 150 and 180 °C. Eur J Wood Wood Prod, 72:33–41.

Repellin V, Guyonnet R (2005) Evaluation of heat-treated wood swelling by differential scanning calorimetry in relation to chemical composition. Holzforschung, 59:28–34.

Ringman R, Pilgård A, Kölle M, Brischke C, Richter K (2016) Effects of thermal modification on Postia placenta wood degradation dynamics: measurements of mass loss, structural integrity and gene expression. Wood Sci Technol, 50:385–397.

Roffael E, Rauch W (1974) Extractives of oak and their influence on the gluing with alkaline phenolic-formaldehyde-resins. Holz Roh Werkst, 32:182–187.

Rowell RM, Ellis WD (1978) Determination of dimensional stabilization of wood by the water-soak method. Wood Fiber Sci, 10:104–111.

Rowell RM, Pettersen, R., Han, J. S., Rowell JS, Tshabalala MA (2005) Cell wall chemistry. In: Rowell RM (ed.) Handbook of wood chemistry and wood composites. CRC Press, Boca Raton.

Runkel RH (1954) Studien über die Sorption der Holzfaser – Erste Mitteilung: Die Sorption der Holzfaser in morphologisch-chemischer Betrachtung. Holz Roh Werkst, 12:226–232.

Runkel RH, Lüthgens M (1956) Untersuchungen über die Heterogenität der Wassersorption der chemischen und morphologischen Komponenten verholzter Zellwände. Holz Roh Werkst, 14:424–441.

Sailer M, Rapp AO, Leithoff H (2000) Improved resistance of Scots pine and spruce by application of an oil-heat treatment. The International Research Group on Wood Preservation, Doc. IRG/WP 00-40162.

Salmén L (2015) Wood morphology and properties from molecular perspectives. Ann For Sci, 72:679–684.

Salmén L, Burgert I (2009) Cell wall features with regard to mechanical performance. A review. Special issue of COST Action E35 2004–2008: Wood machining – micromechanics and fracture. Holzforschung, 63:121-129.

Salmén L, Fahlén J (2006) Reflections on the ultrastructure of softwood fibers. Cell Chem Technol, 40:181–185.

Salmén L, Olsson AM (1998) Interaction between hemicelluloses, lignin and cellulose: Structure-property relationships. J Pulp Pap Sci, 24:99–103.

Salmén L, Olsson A-M, Stevanic JS, Simonović J, Radotić K (2012) Structural organization of the wood polymers in the wood fibre structure. BioResources, 7:521–532.

Sandberg D (1996) The influence of pith and juvenile wood on proportion of cracks in sawn timber when kiln dried and exposed to wetting cycles. Holz Roh Werkst, 54:152–152.

Sandberg D (1997) Radially sawn timber. Holz Roh Werkst, 55:175–182.

Sandberg D, Haller P, Navi P (2013) Thermo-hydro and thermo-hydro-mechanical wood processing: An opportunity for future environmentally friendly wood products. Wood Mater Sci Eng, 8:64–88.

Sandberg D, Söderström O (2006) Crack formation due to weathering of radial and tangential sections of pine and spruce. Wood Mater Sci Eng, 1:12–20.

Sandermann W, Augustin H (1963a) Chemische Untersuchungen über die thermische Zersetzung von Holz – Zweite Mitteilung: Untersuchungen mit Hilfe der Differential-Thermo-Analyse. Holz Roh Werkst, 21:305–315.

Sandermann W, Augustin H (1963b) Chemische Untersuchungen über die thermische Zersetzung von Holz – Erste Mitteilung: Stand der Forschung. Holz Roh Werkst, 21:256–265.

Sandermann W, Augustin H (1964) Chemische Untersuchungen über die thermische Zersetzung von Holz – Dritte Mitteilung: Chemische Untersuchung des Zersetzungsablaufs. Holz Roh Werkst, 22:377–386.

Scheiding W (2016) TMT im Jahr 2016 - ein Update. In: Proceedings of the 9th European TMT Workshop, pp. 12-18, Dresden, Germany.

Schniewind AP (1963) Mechanism of check formation. Forest Prod J, 13:475–480.

Scholz G, Krause A, Militz H (2009) Capillary water uptake and mechanical properties of wax soaked Scots pine. In: Englund F, Hill CAS, Militz H, Segerholm BK (eds.) Proceedings of the 4th European Conference on Wood Modification, pp. 209-212, Stockholm, Sweden.

Schulgasser K, Witztum A (2015) How the relationship between density and shrinkage of wood depends on its microstructure. Wood Sci Technol, 49:389–401.

Shen H, Cao J, Sun W, Peng Y (2016) Influence of post-extraction on photostability of thermally modified Scots pine wood during artificial weathering. Bioresources, 11:4512-4525.

Seborg CO, Stamm AJ (1931) Sorption of water vapor by paper-making materials I – Effect of beating. Ind Eng Chem, 23:1271–1275.

Seborg M, Tarkow H, Stamm AJ (1953) Effect of heat upon the dimensional stabilization of wood. Journal of the Forest Products Research Society, 3:59–67.

Sehlstedt-Persson M, Johansson D, Morén T (2006) Effect of heat treatment on the microstructure of pine, spruce and birch and the influence on capillary absorption. In:

Kurjatko S, Kúdela J, Lagaňa (eds.) Proceedings of the 5th IUFRO Symposium Wood Structure and Wood Properties, pp. 373-379, Sliač, Slovenia.

Sernek M, Boonstra M, Pizzi A, Despres A, Gérardin P (2008) Bonding performance of heat treated wood with structural adhesives. Holz Roh Werkst, 66:173-180.

Singleton VL, Rossi JA (1965) Colorimetry of total phenolics with phosphomolybdic-phosphotungstic acid reagents. Am J Enol Viticult, 16:144–158.

Sivonen H, Maunu SL, Sundholm F, Jämsä S, Viitaniemi P (2002) Magnetic resonance studies of thermally modified wood. Holzforschung, 56:648–654.

Skaar C (1988) Wood-water relations. Springer-Verlag, Berlin

Srebotnik E, Messner K, Foisner R (1988) Penetrability of white rot-degraded pine wood by the lignin peroxidase of *Phanerochaete chrysosporium*. Appl Environ Microb, 54:2608–2614.

Stamm A, Woodruff S (1941) Convenient six-tube vapor sorption apparatus. Ind Eng Chem Anal Ed, 13:836–838.

Stamm AJ (1934) Effect of inorganic salts upon the swelling and the shrinking of wood. J Am Chem Soc, 56:1195–1204.

Stamm AJ (1956) Thermal degradation of wood and cellulose. Ind Eng Chem, 48:413–417.

Stamm AJ, Burr HK, Kline AA (1946) Staybwood – heat-stabilized wood. Ind Eng Chem, 38:630–634.

Stamm AJ, Hansen LA (1937) Minimizing wood shrinkage and swelling – Effect of heating in various gases. Ind Eng Chem, 29:831–833.

Stefke B, Dunky M (2006) Catalytic influence of wood on the hardening behavior of formaldehyde-based resin adhesives used for wood-based panels. J Adhes Sci Technol, 20:761–785.

Stehr M, Johansson I (2000) Weak boundary layers on wood surfaces. J Adhes Sci Technol, 14:1211–1224.

Stirling R, Temiz A (2014) Fungicides and insecticides used in wood preservation. In: Schultz TP, Goodell B, Nicholas DD (eds.) Deterioration and protection of sustainable biomaterials. ACS Symposium Series 1158, Oxford University Press.

Suchy M, Kontturi E, Vuorinen T (2010a) Impact of drying on wood ultrastructure: similarities in cell wall alteration between native wood and isolated wood-based fibers. Biomacromolecules, 11:2161–2168.

Suchy M, Virtanen J, Kontturi E, Vuorinen T (2010b) Impact of drying on wood ultrastructure observed by deuterium exchange and photoacoustic FT-IR spectroscopy. Biomacromolecules, 11:515–520.

Sundqvist B, Karlsson O, Westermark U (2006) Determination of formic-acid and acetic acid concentrations formed during hydrothermal treatment of birch wood and its relation to colour, strength and hardness. Wood Sci Technol, 40:549–561.

Šušteršic Ž, Mohareb A, Chaouch M, Pétrissans M, Petrič M, Gérardin P (2010) Prediction of the decay resistance of heat treated wood on the basis of its elemental composition. Polym Degrad Stabil, 95:94–97.

Sweet MS, Winandy JE (1999) Influence of degree of polymerization of cellulose and hemicellulose on strength loss in fire-retardant-treated Southern pine. Holzforschung, 53:311–317.

Syrjänen T, Kangas E (2000) Heat treated timber in Finland. The International Research Group on Wood Preservation, Doc. IRG/WP 00-40158.

Tarkow H, Stamm AJ (1953) Effect of formaldehyde treatments upon the dimensional stability of wood. Journal of the Forest Products Research Society, 3:33–37.

Thybring EE (2013) The decay resistance of modified wood influenced by moisture exclusion and swelling reduction. Int Biodeter Biodegr, 82:87–95.

Thygesen L G, Engelund ET, Hoffmeyer P (2010) Water sorption in wood and modified wood at high values of relative humidity. Part I: Results for untreated, acetylated, and furfurylated Norway spruce. Holzforschung, 64:315-323.

Tiemann HD (1917) Effect of different methods of drying upon the strength and the hygroscopicity of wood. In: Tiemann HD (ed) The kiln-drying of lumber - A practical and theoretical treatise. pp 256–264, J. B. Lippincott Company, Philadelphia and London.

Tjeerdsma B, Stevens M, Militz H, Van Acker J (2002) Effect of process conditions on moisture content and decay resistance of hydro-thermally treated wood. Holzforsch Holzverw, 54:94–99.

Tjeerdsma BF, Boonstra M, Pizzi A, Tekely P, Militz H (1998) Characterisation of thermally modified wood: molecular reasons for wood performance improvement. Holz Roh Werkst, 56:149–153.

Tjeerdsma BF, Militz H (2005) Chemical changes in hydrothermal treated wood: FTIR analysis of combined hydrothermal and dry heat-treated wood. Holz Roh Werkst, 63:102–111.

Torniainen P, Dagbro O, Morén T (2011) Thermal modification of birch using saturated and superheated steam. In: Larnøy E, Alfredsen G (eds.) Proceedings of the 7th meeting of the Nordic-Baltic Network in Wood Material Science & Engineering (WSE), pp. 43-48, Oslo, Norway.

Townsend T, Dubey B, Tolaymat T, Solo-Gabriele H (2005) Preservative leaching from weathered CCA-treated wood. J Environ Manage, 75:105–113.

Tsoumis GT (2009) Science and technology of wood: structure, properties, utilization. Kessel, Remagen-Oberwinter

Turkulin H, Sell J (2002) Investigations into the photodegradation of wood using microtensile testing - Part 4: Tensile properties and fractography of weathered wood. Holz Roh Werkst, 60:96–105.

UNECE/FAO (2015) Forest products annual market review 2014-2015. Food and Agriculture Organization of the United Nations, United Nations publication, Geneva, Switzerland.

Van Acker J, Michon S, De Boever L, De Windt I, Van den Bulcke J, Van Swaay B, Stevens M (2010) High qualitiy thermal thermal treatment using vacuum based technology to come to more homogeneous durability. In: Hill CAS, Militz H, Andersons B (eds) Proceedings of the 5th European Conference on Wood Modification, pp. 107–118, Riga, Latvia.

Viitaniemi P, Jämsä S, Viitanen H (1997) Method for improving biodegradation resistance and dimensional stability of cellulosic products. Patent, US 5678324 A.

Voß A, Willeitner H (1993) Possibility and problems of characterizing treated wood after service with regard to disposal. The International Research Group on Wood Preservation, Doc. IRG/WP 93-5006.

Voß A, Willeitner H (1995) Characteristics and quantity of impregnated wood waste in Germany. The International Research Group on Wood Preservation, Doc. IRG/WP 95-50041.

Vrentas JS, Vrentas CM (1991) Sorption in glassy polymers. Macromolecules, 24:2404–2412.

Wålinder MEP (2002) Study of lewis acid-base properties of wood by contact angle analysis. Holzforschung, 56:363–371.

Wang J, Cooper P (2005) Effect of oil type, temperature and time on moisture properties of hot oil-treated wood. Holz Roh Werkst, 63:417–422.

Wangaard F, Granados L (1967) The effect of extractives on water-vapor sorption by wood. Wood Sci Technol, 1:253–277.

Ward JC, Simpson WT (2001) Drying defects. In: Dry kiln operator's manual. USDA Agriculture Handbook AH-188

Weaver JW, Nielson JF, Goldstein IS (1960) Dimensional stabilization of wood with aldehydes & related compounds. Forest Prod J, 10:306–310.

Weiland JJ, Guyonnet R (2003) Study of chemical modifications and fungi degradation of thermally modified wood using DRIFT spectroscopy. Holz Roh Werkst, 61:216–220.

Wellons JD, Krahmer RL, Raymond R, Sleet G (1977) Durability of exterior siding plywood with southeast Asian hardwood veneers. Forest Prod J, 27:38–44.

Welzbacher C (2010) TMT-interlab-test to establish suitable quality control techniques - Structure and first results. The International Research Group on Wood Protection, Doc. IRG/WP 10-40503.

Welzbacher C, Brischke C, Rapp A (2007) Influence of treatment temperature and duration on selected biological, mechanical, physical and optical properties of thermally modified timber. Wood Mater Sci Eng, 2:66–76.

Welzbacher C, Rapp A (2002) Comparison of thermally modified wood originating from four industrial scale processes - durability. The International Research Group on Wood Preservation, IRG/WP 02-40229.

Welzbacher CR, Rapp AO (2007) Durability of thermally modified timber from industrial-scale processes in different use classes: Results from laboratory and field tests. Wood Mater Sci Eng, 2:4–14.

Wetzig M, Niemz P, Sieverts T, Bergemann H (2012) Mechanische und physikalische Eigenschaften von mit dem Vakuumpress-Trocknungsverfahren thermisch behandeltem Holz. Bauphysik, 34:1–10.

Whistler RL, Chen CC (1991) Hemicelluloses. In: Lewin M, Goldstein IS (eds) Wood structure and composition. Marcel Decker, Inc., New York.

Wickholm K, Larsson PT, Iversen T (1998) Assignment of non-crystalline forms in cellulose I by CP/MAS ^{13}C NMR spectroscopy. Carbohydr Res, 312:123–129.

Wienhaus O (1999) Modifizierung des Holzes durch eine milde Pyrolyse - abgeleitet aus den allgemeinen Prinzipien der Thermolyse des Holzes. Wissenschaftliche Zeitschrift der Technischen Universität Dresden, 48:17–22.

Wikberg H, Maunu S (2004) Characterisation of thermally modified hard- and softwoods by ^{13}C CPMAS NMR. Carbohydr Polym, 58:461–466.

Willeitner H (1973) Polution in wood preservation - aspects and problems. The International Research Group on Wood Preservation, Doc. IRG/WP/55.

Willems W (2009) A novel economic large-scale production technology for high-quality thermally modified wood. In: Englund F, Hill CAS, Militz H, Segerholm BK (eds.) Proceedings of the 4th European Conference on Wood Modification, pp. 31–35, Stockholm, Sweden.

Willems W (2014a) The water vapor sorption mechanism and its hysteresis in wood: the water/void mixture postulate. Wood Sci Technol, 48:499–518.

Willems W (2014b) Hydrostatic pressure and temperature dependence of wood moisture sorption isotherms. Wood Sci Technol, 48:483–498.

Willems W (2015) A critical review of the multilayer sorption models and comparison with the sorption site occupancy (SSO) model for wood moisture sorption isotherm analysis. Holzforschung, 69:67–75.

Willems W (2010) FIRMOLIN - die schonende thermische Holzmodifikation. In: Proceedings of the 6th European TMT-Workshop, Dresden, Germany.

Willems W, Altgen M, Militz H (2015a) Comparison of EMC and durability of heat treated wood from high versus low water vapour pressure reactor systems. International Wood Products Journal, 6:21–26.

Willems W, Lykidis C, Altgen M, Clauder L (2015b) Quality control methods for thermally modified wood. Holzforschung, 69:875–884.

Willems W, Mai C, Militz H (2013) Thermal wood modification chemistry analysed using van Krevelen's representation. International Wood Products Journal, 4:166–171.

Willems W, Tausch A, Militz H (2010) Evidence for an antioxidant mechanism in the durability of high-pressure steam modified wood. In: Hill CAS, Militz H, Andersons B (eds) Proceedings of the 5th European Conference on Wood Modification, pp. 127–134, Riga, Latvia.

Williams RS (2005) Weathering of wood. In: Rowell RM (ed) Handbook of wood chemistry and wood composites. Chap. 7, pp. 139–185, CRC Press, Boca Raton.

Winandy JE, Lebow PK (2001) Modeling strength loss in wood by chemical composition. Part I: An individual component model for Southern pine. Wood Fiber Sci, 33:239–254.

Winandy JE, Rowell RM (2005) Chemistry of wood strength. In: Rowell RM (ed) Handbook of wood chemistry and wood composites. Chap. 11, pp. 303-347, CRC Press, Boca Raton.

Xiao ZF, Xie YJ, Militz H, Mai C (2010) Effects of modification with glutaraldehyde on the mechanical properties of wood. Holzforschung, 64:475–482.

Xie Y, Hill CAS, Sun DY, Jalaludin Z, Wang Q (2012) Effects of extractives on the dynamic water swelling behaviour and fungal resistance of Malaysian hardwood. J Trop For Sci, 24:231–240.

Xie Y, Krause A, Militz H, Turkulin H, Richter K, Mai C (2007) Effect of treatments with 1,3-dimethylol-4,5-dihydroxy-ethyleneurea (DMDHEU) on the tensile properties of wood. Holzforschung, 61:43–50.

Yasuda R, Minato K, Norimoto M (1994) Chemical modification of wood by non-formaldehyde cross-linking reagents. Wood Sci Technol, 28:209–218.

Yildiz S (2002) Effects of heat treatment on water repellence and anti swelling efficiency of beech wood. The International Research Group on Wood Preservation, IRG/WP 02-40223.

Yildiz S, Tomak ED, Yildiz UC, Ustaomer D (2013) Effect of artificial weathering on the properties of heat treated wood. Polym Degrad Stabil, 98:1419–1427.

Zabel RA, Morrell JJ (1992) Wood microbiology. Decay and its prevention. Academic Press, Inc., San Diego

Zaman A, Alén R, Kotilainen R (2000) Thermal behavior of Scots pine (*Pinus sylvestris*) and Silver birch (*Betula pendula*) at 200-230°C. Wood Fiber Sci, 32:138–143.

Zauer M, Pfriem A, Wagenführ A (2013) Toward improved understanding of the cell-wall density and porosity of wood determined by gas pycnometry. Wood Sci Technol, 47:1197–1211.

Zelinka SL, Gleber SC, Vogt S, López Rodríguez GM, Jakes JE (2014) Threshold for ion movements in wood cell walls below fiber saturation observed by X-ray fluorescence microscopy (XFM). Holzforschung, 69:441-448.

Zelinka SL, Lambrecht MJ, Glass SV, Wiedenhoeft AC, Yelle DJ (2012) Examination of water phase transitions in Loblolly pine and cell wall components by differential scanning calorimetry. Thermochim Acta, 533:39–45.

Zelinka SL, Ringman R, Pilgård A, Thybring EE, Jakes JE, Richter K (2016) The role of chemical transport in the brown-rot decay resistance of modified wood. International Wood Products Journal, 7:66–70.

Zhang J, Kamdem DP (2000) FTIR Characterization of copper ethanolamine-wood interaction for wood preservation. Holzforschung, 54:119-122.

Curriculum vitae

Name: Michael Altgen

Born: 04.03.1985 in Mechernich, Germany

Nationality: German

Work experience

Scientific assistant   2011-2016

Wood Biology and Wood Products, Burckhardt Institute, Faculty of Forest Sciences and Forest Ecology, Georg-August University, Göttingen, Germany

Education

PhD student   2011-2016

Wood Biology and Wood Products, Burckhardt Institute, Faculty of Forest Sciences and Forest Ecology, Georg-August University, Göttingen, Germany

Master of Science   2008-2011
Forest Sciences and Forest Ecology, Georg-August University, Göttingen, Germany

Bachelor of Science   2005-2008
Forest Sciences and Forest Ecology, Georg-August University, Göttingen, Germany